Therapeutic Landscapes in Colorectal Carcinoma

Therapeutic Landscapes in Colorectal Carcinoma

Editors

Antonio M Scanu
Maria Rosaria De Miglio

MDPI

Basel • Beijing • Wuhan • Barcelona • Belgrade • Novi Sad • Cluj • Manchester

Editors

Antonio M Scanu
Medicine, Surgery
and Pharmacy
University of Sassari
Sassari
Italy

Maria Rosaria De Miglio
Medicine, Surgery
and Pharmacy
University of Sassari
Sassari
Italy

Editorial Office
MDPI
St. Alban-Anlage 66
4052 Basel, Switzerland

This is a reprint of articles from the Special Issue published online in the open access journal *Medicina* (ISSN 1648-9144) (available at: www.mdpi.com/journal/medicina/special_issues/Colorectal_Oncology).

For citation purposes, cite each article independently as indicated on the article page online and as indicated below:

Lastname, A.A.; Lastname, B.B. Article Title. *Journal Name* **Year**, *Volume Number*, Page Range.

ISBN 978-3-0365-9441-5 (Hbk)
ISBN 978-3-0365-9440-8 (PDF)
doi.org/10.3390/books978-3-0365-9440-8

Contents

About the Editors

Antonio M Scanu

Professor Antonio Mario Scanu is an associate professor of general surgery at the Department of Medicine, Surgery, and Pharmacy of the University of Sassari, Italy. He obtained his degree in 1991 from the University of Sassari, his postgraduate degree in general surgery in 1996, and his postgraduate degree in digestive system surgery and surgical digestive endoscopy in 2002. He has been a researcher in general surgery and lecturer since 1995, and an associate professor since 2002. In recognition of his merits, he received funding from the Italian scientific research award system (PRIN 2006) in 2006.

He is a member of the Italian Society of Surgery (SIC), the Italian Society of Hospital Surgeons (ACOI), and other surgical scientific societies.

He is an associate editor of *Frontiers in Oncology*. He has considerable experience in the fields of colorectal surgery (CRC and IBD) and translational surgery (particular interest in the study of genomic and epigenomic deregulations in colorectal cancer). He focused his research on the integration of multi-omics approaches for the study of CRC, performing transcriptome analysis studies using innovative massive sequencing approaches, and also identifying epigenetic deregulations involved in tumor pathogenesis.

Maria Rosaria De Miglio

Professor De Miglio is an associate professor of general pathology at the Department of Medicine, Surgery, and Pharmacy of the University of Sassari, Italy. She achieved her M.D. in 1992 at the University of Pisa, her Ph.D. in 1997 at the University of Turin, and her post-doctoral experience in 1998 at the International Agency for Research on Cancer, "Unit of Genetic Cancer Susceptibility", Lyon, France. In recognition of her achievements, she received awards and recognition for the scientific system of rewards for scientific research from the Sardinian Region in 2011 and, in 2016, the award for best poster for scientific research at the 48th National Meeting of SIBioC. She is a member of the American Association for Cancer Research, the Italian Society of Pathology and Translational Medicine, and the Italian Society for the Study of Connective Tissues. She is the associate editor of *Frontiers in Oncology, Scientific Reports, Medicina*, and *Technology in Cancer Research and Treatment*. She has remarkable experience in the field of oncology, with particular interest in the study of genomic and epigenomic deregulations in colorectal carcinoma and triple-negative breast cancer, which showed strong aggressiveness and therapeutic options that were not very effective. She has focused her searches on the integration of multi-omics approaches for the study of CRC, performing transcriptome analysis studies using innovative massive sequencing approaches, and also identifying epigenetic deregulations involved in tumor pathogenesis.

medicina

MDPI

Editorial

Therapeutic Landscapes in Colorectal Carcinoma

Antonio Mario Scanu and Maria Rosaria De Miglio *

Department of Medicine, Surgery and Pharmacy, University of Sassari, 07100 Sassari, Italy; scanu@uniss.it
* Correspondence: demiglio@uniss.it; Tel.: +39-079-228016

Colorectal cancer (CRC) is a disease of major public health and socioeconomic concern. According to GLOBOCAN 2020, in terms of incidence and mortality rates, CRC classifies third and second, respectively; in 2020, over 1.9 million novel CRC cases and 935.000 deaths were reported to occur, accounting to 1 in 10 cancer cases and deaths [1]. Recently, a decreased CRC incidence in high-incidence countries has been reported, which might be ascribed to population-level changes regarding healthier lifestyle choices, as well as to increased colonoscopy screening and an improvement in therapies [1]. However, more recurrent favorable prognosis affecting adults aged ≥50 years hides the increasing rates of early-onset CRC and age at diagnosis <50 years in various countries. In 2018, in fact, to alleviate the increasing burden of early-onset CRC, the recommended age for screening initiation for individuals at average risk has been lowered from 50 to 45 years by the American Cancer Society [2]. In October 2020, the US Preventive Services Task Force issued a recommendation statement (https://uspreventiveservicestaskforce.org/uspstf/recommendation/colorectal-cancer-screening, accessed on 18 May 2021).

Almost 20% of CRC patients show metastases at diagnosis, and metastatic CRC (mCRC) is most often a non-curable disease [3]. The five-year survival rate is 90% at stage I, severely decreasing to around 10% at stage IV [4]. Due to the high frequency of metastases and drug resistance, it is still one of the hard-to-treat cancers, despite all the advances in CRC biology knowledge and therapeutic improvements. Given that CRC is a heterogeneous and multifactorial disease with extremely different prognoses and responses to treatment, it is crucial to understand specific pathway abnormalities in order to improve diagnosis, prognosis, and therapeutic approaches. The ability to identify specific biomarkers and detect unequivocal molecular targets associated with early cancer signs will provide valuable support to achieve new targeted therapies and decrease CRC mortality rates. Even though different crucial genes and pathways have already been associated with CRC biology, the prognostic and predictive roles of many of these genomic alterations are unknown and have no influence on treatment decisions in metastatic patients. Recently, the identification of KRAS and NRAS gene mutations has been adopted as an extensively recognized molecular test in mCRC clinical treatment [5]. In order to identify patients with a more aggressive clinical outcome, BRAF V600E mutation has been approved as a prognostic biomarker [5]. Moreover, a relationship between MSI-high CRC and an effective response to the immune checkpoint blockade through anti-PD1 therapy has been proven in mCRC patients [6].

Accordingly, interpreting more effectively available data and further investigating molecular mechanisms triggering CRC pathogenesis are vital steps in order to achieve a higher level of prevention, outcome, and therapy in patients affected by CRC.

This Special Issue, entitled "Therapeutic Landscapes in Colorectal Carcinoma", provides us with 11 really opportune articles, 9 original articles and 2 reviews, which may provide us with a deeper insight into the latest advances involving knowledge on CRC from scientific, translational, and clinical points of view. Through this Special Issue, our main purpose was to shed light on some state-of-the-art research on CRC. Here, the reader will find papers on prognostic factors, as well as on responses to neoadjuvant and adjuvant

Citation: Scanu, A.M.; De Miglio, M.R. Therapeutic Landscapes in Colorectal Carcinoma. *Medicina* **2023**, *59*, 821. https://doi.org/10.3390/medicina59050821

Received: 13 March 2023
Revised: 4 April 2023
Accepted: 21 April 2023
Published: 23 April 2023

CRC treatments. Finally, we discuss the research trends and hotspots for CRC therapies that emerged during the COVID-19 pandemic, such as the potential achievement of a transition from CRC clinical research to precision medicine, together with a special emphasis given to new single-cell-based techniques.

The most important cause of death in CRC is disease progression due to metastasis and drug resistance. Therefore, the identification of unambiguous molecular biomarkers to predict disease aggressiveness and drug response is needed. Interestingly, it has been observed that a higher expression of RIPK2, which acts as a critical mediator necessary in different immune and inflammatory pathways, was associated with a high expression of VEGFA and increased mortality in CRC patients, suggesting its promising role as a prognostic tool [7].

Neoadjuvant chemotherapy (NAC) for locally advanced CRC has become progressively more commonly administered in clinical settings; however, its applicability is currently under debate. Zeng et al. demonstrated that a combination of NAC and adjuvant chemotherapy was not only safe, but also caused a noteworthy reduction in the primary tumor size and stage; in addition, in a lower percentage of patients, a complete pathological response (pT0) was observed. Eventually, long-term outcomes were similar to results in patients who after diagnosis directly underwent surgery.

CRC can be distinguished as mismatch repair-deficient (dMMR), with high levels of microsatellite instability (MSI-H), mismatch repair proficiency (pMMR), and microsatellite stability (MSS). Approximately 15–30% of CRC patients are affected by MSI-H [8], which has emerged as an important predictor of sensitivity to immune checkpoint inhibitors (ICIs). Considering that the European Medicines Agency (EMA) has recently approved pembrolizumab as the first ICI in the treatment of dMMR–MSI-H mCRC, Lungulescu et al. assessed the regional variability of MSI-H CRC in Romania, observing a higher regional MSI-H prevalence (21%) compared to the literature; they suggested that analyzing both geographical variations and clinical features in CRC patients is essential as advanced therapies, diagnostic tools, and innovative methods of treatments delivery are regularly being developed [9].

For instance, the study of patients' clinical course after the diagnosis of de novo CRC after liver transplantation, with an emphasis on the influence of immunosuppressive management, showed a significantly enhanced survival rate when the immunosuppressive therapy was reduced in an individualized manner, leading to an optimal oncological therapy and higher survival rates [10].

Chen et al. conducted a nationwide, large-scale, retrospective cohort study to compare the effectiveness of adjuvant chemotherapy based on uracil and tegafur (UFT) in patients with stage II CRC; the study showed that, in the 15-year follow-up cohorts, UFT did not induce differences compared to the observation group in terms of disease-free survival (DFS) and overall survival (OS) rates. However, DFS notably increased in patients with stage IIA CRC treated with UFT as postoperative adjuvant chemotherapy compared with DFS in the observation group [11].

Seeing the scarce responses to standard systemic treatment in BRAF-mutated mCRC patients, through a single-center case series that included patients with BRAF-mutated mCRC in Asia, Yeh et al. analyzed the real effects of triplet therapy (dabrafenib, trametinib, and panitumumab) after previous systemic treatment failure. An adequate safety profile and acceptable treatment efficacy resulted from the study. Moreover, an interestingly higher OS was discovered in patients with left-sided mCRC than in patients with right-sided tumor [12].

Considering that the response to BRAF inhibitors of BRAF-mutated mCRC is rather brief, and progression is the rule, through an in silico study, Voutsadakis et al. suggested that targeted therapies for CRC showing BRAF mutations with or without PIK3CA mutations can be improved on the basis of the global molecular environment of these disease. The results showed that CRCs with BRAF mutations, and with or without PIK3CA mutations, vary in their MSI status and mimic CRC tissues with APC and TP53 mutations.

CTNNB1, WRN, and CAD affected OS. Additionally, BRAF inhibitor sensitivity in CRC cell lines is shown by SACS mutations and PRKN loss [13].

Patients with mCRC show a poor prognosis despite the therapeutic options currently available. Regorafenib is an oral tyrosine kinase inhibitor that can be administered to treat refractory mCRC; however, no effective predictive markers for regorafenib treatment have been identified yet. De Summa et al. assessed somatic mutations of genes involved in immunological and inflammatory responses using an NGS platform to identify potential biomarkers in mCRC patients long and short responded to regorafenib. These results underline the presence of mutations in TGFBR1, TGFBR2, and TGFBR3 genes, suggesting the role of the TGF-β pattern in a prolonged response to the drug [14].

Voutsadakis et al. analyzed the therapeutic implications of 20q11.21 amplification in 12 CRC cell lines. Amplified 20q11.21 cell lines are sensible to different tyrosine kinase inhibitors and no-responders to drugs targeting the mitotic apparatus and microtubules. CRISPR and RNAi dependency analysis identified YAP1 and JUP as recurrent gene dependencies in cell lines. Therefore, amplified 20q11.21 gene cell line models of CRC with resistance or sensitivity to various drug categories could be adopted within in vitro models to favor clinical drug development in this tumor [15].

CRC shows heterogeneous genomic, epigenomic, and transcriptomic aberrations. Intra-tumoral heterogeneity (ITH) can be observed within a tumor in which cancer cell subpopulations with diverse genomic characters exist in a patient. As reported by Angius et al., ITH analysis is a promising new frontier that lays the basis toward effective CRC diagnosis and treatment. Genome and transcriptome sequencing, together with editing technologies, are transforming biomedical research, and represent the most encouraging tools for defeating unmet clinical and research challenges. Bulk and single-cell next-generation sequencing are recognizing genomic and transcriptional heterogeneity in primary and metastatic tumors [16].

Finally, Kopel et al. discussed research trends and hotspots for CRC management in the period of COVID-19 pandemic emergency. The authors suggested that the COVID-19 pandemic caused the world to pause and adopt lockdown measures that block CRC screening programs, which resulted in a dramatic increase in late-stage CRC cases and a general loss of life years due to the lack of appropriate treatments for CRC patients [17].

This Special Issue, titled "Therapeutic Landscapes in Colorectal Carcinoma", illustrates a stimulating collection of articles written by international specialists that can promote discussions and ideas among colleagues working on CRC. We hope that it can encourage translational and interdisciplinary collaborations, leading to a definitive understanding of strategies to overcome and inhibit CRC progression and metastasis.

Author Contributions: Conceptualization, A.M.S. and M.R.D.M.; writing—original draft preparation, M.R.D.M.; writing—review and editing, A.M.S. and M.R.D.M. All authors have read and agreed to the published version of the manuscript.

Conflicts of Interest: The authors declare no conflict of interest.

References

1. Sung, H.; Ferlay, J.; Siegel, R.L.; Laversanne, M.; Soerjomataram, I.; Jemal, A.; Bray, F. Global Cancer Statistics 2020: GLOBOCAN Estimates of Incidence and Mortality Worldwide for 36 Cancers in 185 Countries. *CA Cancer J. Clin.* **2021**, *71*, 209–249. [CrossRef] [PubMed]
2. Wolf, A.; Fontham, E.T.; Church, T.R.; Flowers, C.R.; Guerra, C.E.; LaMonte, S.J.; Etzioni, R.; McKenna, M.T.; Oeffinger, K.C.; Tina Shih, Y.-C.; et al. Colorectal Cancer Screening for Average-Risk Adults: 2018 Guideline Update From the American Cancer Society; 6 Independent Retired Physician and Patient Advocate. *CA A Cancer J. Clin.* **2018**, *68*, 250–281. [CrossRef] [PubMed]
3. Riihimäki, M.; Hemminki, A.; Sundquist, J.; Hemminki, K. Patterns of Metastasis in Colon and Rectal Cancer. *Sci. Rep.* **2016**, *6*, 29765. [CrossRef] [PubMed]
4. Brenner, H.; Kloor, M.; Pox, C.P. Colorectal Cancer. *Lancet* **2014**, *383*, 1490–1502. [CrossRef] [PubMed]
5. Van Cutsem, E.; Köhne, C.H.; Láng, I.; Folprecht, G.; Nowacki, M.P.; Cascinu, S.; Shchepotin, I.; Maurel, J.; Cunningham, D.; Tejpar, S.; et al. Cetuximab plus Irinotecan, Fluorouracil, and Leucovorin as First-Line Treatment for Metastatic Colorectal Cancer:

Updated Analysis of Overall Survival According to Tumor KRAS and BRAF Mutation Status. *J. Clin. Oncol.* **2011**, *29*, 2011–2019. [CrossRef] [PubMed]
6. Le, D.T.; Durham, J.N.; Smith, K.N.; Wang, H.; Bartlett, B.R.; Aulakh, L.K.; Lu, S.; Kemberling, H.; Wilt, C.; Luber, B.S.; et al. Mismatch Repair Deficiency Predicts Response of Solid Tumors to PD-1 Blockade. *Science* **2017**, *357*, 409–413. [CrossRef] [PubMed]
7. Jaafar, R.F.; Ibrahim, Z.; Ataya, K.; Hassanieh, J.; Ard, N.; Faraj, W. Receptor-Interacting Serine/Threonine-Protein Kinase-2 as a Potential Prognostic Factor in Colorectal Cancer. *Medicina* **2021**, *57*, 709. [CrossRef] [PubMed]
8. Hause, R.J.; Pritchard, C.C.; Shendure, J.; Salipante, S.J. Classification and Characterization of Microsatellite Instability across 18 Cancer Types. *Nat. Med.* **2016**, *22*, 1342–1350. [CrossRef] [PubMed]
9. Lungulescu, C.; Croitoru, V.M.; Volovat, S.R.; Cazacu, I.M.; Turcu-Stiolica, A.; Gheonea, D.I.; Sur, D.; Lungulescu, C.V. An Insight into Deficient Mismatch Repair Colorectal Cancer Screening in a Romanian Population—A Bi-Institutional Pilot Study. *Medicina* **2021**, *57*, 847. [CrossRef] [PubMed]
10. Ossami Saidy, R.R.; Wegener, E.; Uluk, D.; Dittrich, L.; Schöning, W.; Lurje, G.; Öllinger, R.; Modest, D.P.; Tacke, F.; Haase, O.; et al. A Reduction of Calcineurin Inhibitors May Improve Survival in Patients with De Novo Colorectal Cancer after Liver Transplantation. *Medicina* **2022**, *58*, 1755. [CrossRef] [PubMed]
11. Chen, P.-H.; Jhou, H.-J.; Chung, C.-H.; Wu, Y.-Y.; Huang, T.-C.; Lee, C.-H.; Chien, W.-C.; Chen, J.-H. Benefit of Uracil–Tegafur Used as a Postoperative Adjuvant Chemotherapy for Stage IIA Colon Cancer. *Medicina* **2022**, *59*, 10. [CrossRef] [PubMed]
12. Yeh, J.H.; Tsai, H.L.; Chen, Y.C.; Li, C.C.; Huang, C.W.; Chang, T.K.; Su, W.C.; Chen, P.J.; Liu, Y.P.; Wang, J.Y. BRAF, MEK, and EGFR Triplet Inhibitors as Salvage Therapy in BRAF-Mutated Metastatic Colorectal Cancer&emdash;A Case Series Study Target Therapy of BRAF-Mutated MCRC. *Medicina* **2021**, *57*, 1339. [CrossRef] [PubMed]
13. Voutsadakis, I.A. Sensitivities and Dependencies of BRAF Mutant Colorectal Cancer Cell Lines with or without PIK3CA Mutations for Discovery of Vulnerabilities with Therapeutic Potential. *Medicina* **2022**, *58*, 1498. [CrossRef] [PubMed]
14. De Summa, S.; Danza, K.; Pilato, B.; Matera, G.; Fasano, R.; Calabrese, A.; Lacalamita, R.; Silvestris, N.; Tommasi, S.; Argentiero, A.; et al. A Promising Role of TGF-β Pathway in Response to Regorafenib in Metastatic Colorectal Cancer: A Case Report. *Medicina* **2021**, *57*, 1241. [CrossRef] [PubMed]
15. Voutsadakis, I.A. Cell Models for Chromosome 20q11.21 Amplification and Drug Sensitivities in Colorectal Cancer. *Medicina* **2021**, *57*, 860. [CrossRef] [PubMed]
16. Angius, A.; Scanu, A.M.; Arru, C.; Muroni, M.R.; Carru, C.; Porcu, A.; Cossu-Rocca, P.; De Miglio, M.R. A Portrait of Intratumoral Genomic and Transcriptomic Heterogeneity at Single-Cell Level in Colorectal Cancer. *Medicina* **2021**, *57*, 1257. [CrossRef] [PubMed]
17. Kopel, J.; Ristic, B.; Brower, G.L.; Goyal, H. Global Impact of COVID-19 on Colorectal Cancer Screening: Current Insights and Future Directions. *Medicina* **2022**, *58*, 100. [CrossRef] [PubMed]

medicina

Review

A Portrait of Intratumoral Genomic and Transcriptomic Heterogeneity at Single-Cell Level in Colorectal Cancer

Andrea Angius [1,*], Antonio Mario Scanu [2], Caterina Arru [3], Maria Rosaria Muroni [2], Ciriaco Carru [3], Alberto Porcu [2], Paolo Cossu-Rocca [2] and Maria Rosaria De Miglio [2,*]

1 Istituto di Ricerca Genetica e Biomedica (IRGB), CNR, Cittadella Universitaria di Cagliari, 09042 Monserrato, Italy
2 Department of Medical, Surgical and Experimental Sciences, University of Sassari, Via P. Manzella, 4, 07100 Sassari, Italy; scanu@uniss.it (A.M.S.); mrmuroni@uniss.it (M.R.M.); alberto@uniss.it (A.P.); rocco@uniss.it (P.C.-R.)
3 Department of Biomedical Sciences, University of Sassari, 07100 Sassari, Italy; 30039590@studenti.uniss.it (C.A.); carru@uniss.it (C.C.)
* Correspondence: andrea.angius@irgb.cnr.it (A.A.); demiglio@uniss.it (M.R.D.M.); Tel.: +39-0706754543 (A.A.); +39-079228016 (M.R.D.M.)

Citation: Angius, A.; Scanu, A.M.; Arru, C.; Muroni, M.R.; Carru, C.; Porcu, A.; Cossu-Rocca, P.; De Miglio, M.R. A Portrait of Intratumoral Genomic and Transcriptomic Heterogeneity at Single-Cell Level in Colorectal Cancer. *Medicina* **2021**, *57*, 1257. https://doi.org/10.3390/medicina57111257

Academic Editor: Konstantinos Dimas

Received: 30 September 2021
Accepted: 15 November 2021
Published: 17 November 2021

Abstract: In the study of cancer, omics technologies are supporting the transition from traditional clinical approaches to precision medicine. Intra-tumoral heterogeneity (ITH) is detectable within a single tumor in which cancer cell subpopulations with different genome features coexist in a patient in different tumor areas or may evolve/differ over time. Colorectal carcinoma (CRC) is characterized by heterogeneous features involving genomic, epigenomic, and transcriptomic alterations. The study of ITH is a promising new frontier to lay the foundation towards successful CRC diagnosis and treatment. Genome and transcriptome sequencing together with editing technologies are revolutionizing biomedical research, representing the most promising tools for overcoming unmet clinical and research challenges. Rapid advances in both bulk and single-cell next-generation sequencing (NGS) are identifying primary and metastatic intratumoral genomic and transcriptional heterogeneity. They provide critical insight in the origin and spatiotemporal evolution of genomic clones responsible for early and late therapeutic resistance and relapse. Single-cell technologies can be used to define subpopulations within a known cell type by searching for differential gene expression within the cell population of interest and/or effectively isolating signal from rare cell populations that would not be detectable by other methods. Each single-cell sequencing analysis is driven by clustering of cells based on their differentially expressed genes. Genes that drive clustering can be used as unique markers for a specific cell population. In this review we analyzed, starting from published data, the possible achievement of a transition from clinical CRC research to precision medicine with an emphasis on new single-cell based techniques; at the same time, we focused on all approaches and issues related to this promising technology. This transition might enable noninvasive screening for early diagnosis, individualized prediction of therapeutic response, and discovery of additional novel drug targets.

Keywords: colorectal carcinoma; intratumor heterogeneity; single-cell next-generation sequencing; precision medicine

1. Introduction

Genome and transcriptome sequencing and editing technologies, supplemented with machine learning, are setting the stage for the transition from traditional to precision medicine [1–8]. In cancer studies, we are observing a promising transition from research on spatiotemporal tumor heterogeneity [9–11] to early-stage clinical trials [12–17].

As inter-tumor heterogeneity is characterized by variability in patients with the same histologic type [18,19], this might influence clinical care in cancer by providing targeted therapies based on tumor genetic features. We can now monitor clonal dynamics during treatment or identify clinical resistance during disease progression.

Intra-tumoral heterogeneity (ITH) is detectable: subpopulations of cancer cells differ in genome features and tumor areas and/or may evolve/differentiate over time [20–22]. Thus, ITH represents a key determinant of treatment failure, drug resistance, and disease recurrence [19].

Colorectal carcinoma (CRC) is a leading mortality cause worldwide [10,11] and is characterized by heterogeneous genomic, epigenomic and transcriptomic alterations [23–29]. The heterogeneous nature of CRC may also be related to colorectal cancer stem cells (CCSCs): a small population with stem-like behavior responsible for tumor progression, recurrence, and resistance to therapy [16].

CRC treatment has been standardized based on clinicopathological and genetic features (KRAS/NRAS/BRAF mutation and Microsatellite instability (MSI) status), as well as based on tumor staging. Characterization of multiple samples from the same patient proved to be a significant ITH indicator between different areas of the same tumor (spatial heterogeneity) as well as comparing the primary tumor and a subsequent local or distant recurrence (temporal heterogeneity) [18].

The ability of next-generation sequencing (NGS) both at whole genome and single-cell levels to identify disease-associated variants and tumor features triggered a renewed interest on the effectiveness of biomedical and oncology research [1,2,18,30]. Whole genomic and transcriptomic profiling only shows us the average cellular characteristics, thus hiding critical aspects of tumor heterogeneity. Deep bulk sequencing can only capture 1% of the cell population, excluding some types such as circulating tumor cells. Therefore, single-cell techniques allow us to accurately explore cellular properties [31]. Despite recent advances, single-cell next generation sequencing (scNGS) suffers from limited availability of public data/databases and the lack of standardization of laboratory protocols and computer analysis.

Although over the years conventional research has improved, as well as outcomes in CRC patients through diagnosis standardization, staging, and multimodal treatment, important critical and clinical issues remain unresolved [6–9,32].

Recent considerations of dynamic clonal evolution [33], spatiotemporal detection of genomic clones, circulating tumor DNA (ctDNA), identification of ITH [34] and circulating cell heterogeneity [35] allow delineation and improvement of therapeutic failure and relapse [36]. Single-cell transcriptomics, CRISPR-Cas9, and their combination returned exciting data on cell-to-cell drug-dependent variability [9,37,38]. Pioneering combinations of scNGS, CRISPR-Cas, and Hi-C technologies raise high hopes for understanding the linear and nonlinear interactions that control gene expression at single-cell resolution [39].

Based on a review of published data, we aimed at discussing the possible achievement of a transition from CRC clinical research to precision medicine with a special emphasis on new single-cell-based techniques, focusing on all approaches and issues related to these technologies. This transition may provide feasible non-invasive screening procedures for early diagnosis, individualized prediction of therapeutic response and discovery of additional novel drug targets.

2. Innovative Methodologies Applied to Precision Medicine

Proper analysis and extensive use of the large amount of data generated from single scNGS experiments are very challenging and require experienced personnel. A full understanding of the experimental and computational pathways starting from the wet lab to the sophisticated computer analysis of data is needed. Attention must be given to quality control measures for determining which individual cells to include for further examination, data normalization methods, clustering, and visualization for dimensional reduction of data into a two-dimensional graph.

As far as the experimental design is involved, no less significant are the costs that vary from EUR 1–2 to a few cents per cell. The price is highly dependent on the number of cells sequenced, the desired sequencing depth, and the sequencing platform used.

Regardless of cell separation method and labeling of mRNA molecules, all approaches rely on similar computational pipelines for transcriptional profiling. Some concepts are applicable to the majority of single-cell sequencing platforms that use DNA barcodes as an approach to link mRNA transcripts to a single-cell source. Single-cell technologies can be used to define subpopulations within a known cell type by seeking differential gene expression within the cell population of interest; at the same time, they can effectively isolate signal from rare cell populations that would not be detectable by other methods. Each individual cell analysis is driven by clustering of cells based on their differentially expressed genes. The genes driving the clustering can be utilized as unique markers for a specific cell population.

2.1. Generation of Single-Cell Expression Datasets

There are several high-throughput single-cell sequencing platforms on the market at the moment: the most widely used and cost-effective are Fluidigm C1, DropSeq and Chromium 10X [40,41]. These technologies can define the transcriptional profile from hundreds to thousands of individual cells simultaneously. They all are based on labeling mRNA molecules with DNA barcodes during reverse transcription and/or subsequent steps, which allow for indexing of transcripts to their individual cells of origin [42–44].

The various cell capture methods have to consider several parameters that differ from one method to the other and affect the final sequencing results. The main parameters are the number of starting cells (which varies from about 1000 to 500,000), the method of cell separation (cell capture, droplet-based, etc.) and the efficiency of cell capture [45].

The C1 system isolates single cells into individual reaction chambers in the Fluidigm integrated fluidic circuit (IFC). The optically clear IFC enables the operator to automatically stain captured cells and examine them by microscopy for viability, surface markers or reporter genes. Cell lysing, reverse transcription, and cDNA amplification are performed on the C1 Single-Cell Auto Prep IFC using a SMARTer Ultra Low RNA Kit for cDNA synthesis [46–49] followed by a standard Illumina NGS library protocol.

Droplet-based single-cell gene expression approaches, including DropSeq and the 10X platform, use microfluidic chips to isolate single cells along with individual microspheres embedded in oil droplets using a microfluidic so that each droplet contains a single cell [50,51]. The microspheres are coated with DNA oligos that are composed of a poly(T) tail at the 3′ end for capturing cellular mRNA, and at the 5′ end possess a cellular barcode that is identical for each oligo coating a single bead and an individual unique molecular identifier (UMI) barcode for high diversity [52–54]. The transcripts from each individual cell captured and labeled by the DNA oligos attached to a bead are reverse transcribed and amplified with PCR; subsequently, they are sequenced using a high-throughput platform after breaking and pooling droplet contents.

2.2. Bioinformatics Approaches to Single-Cell Analysis

scRNA-seq data analysis poses several unique computational challenges that need the adaptation of existing workflows, as well as the development and application of new analytical strategies (Figure 1). Many analytical procedures rely on specialized algorithms developed and made available to the international community by reference bioinformatics laboratories [55]. Sequencing data from various methods are mostly produced using standard NGS methodology and Illumina instrumentation. They are aligned to a reference genome to annotate each transcript to its gene. Cell barcodes allow computational linkage of each gene transcript to its cell of origin. The number of individual gene transcripts expressed in each cell is counted using UMIs, allowing the assembly of digital gene expression arrays (DGEs), which are tables of cell barcodes and gene counts.

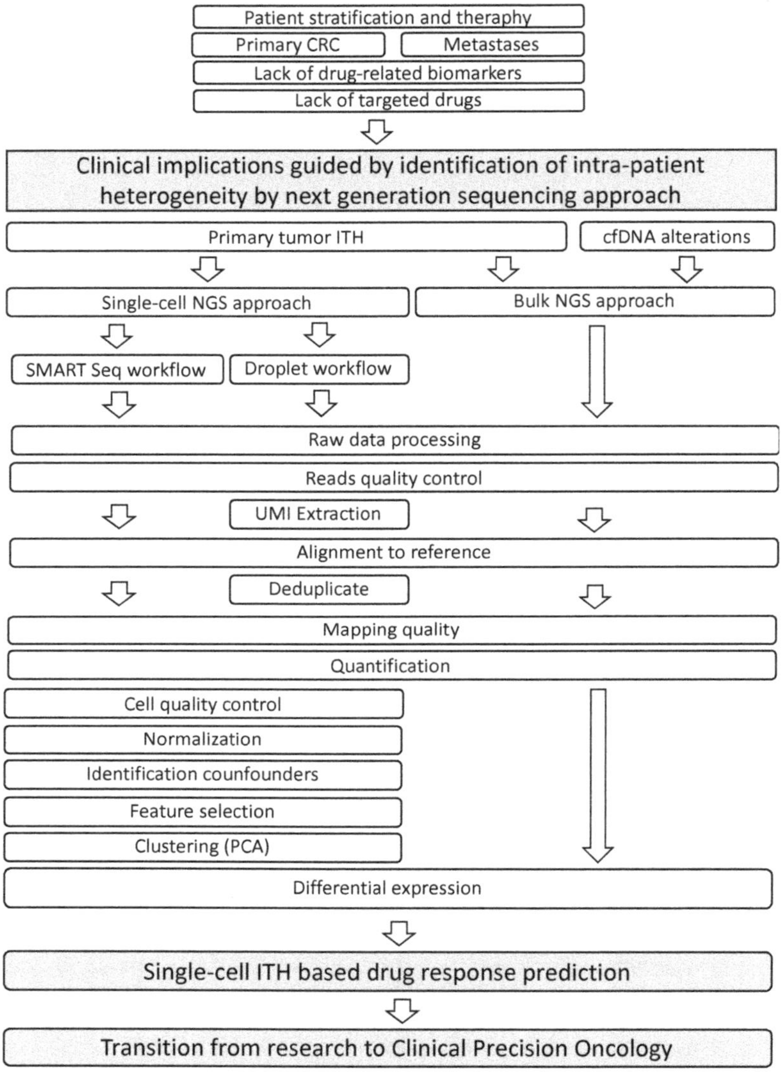

Figure 1. Brief outline of the state of the art of colorectal cancer management, issues to be addressed and potential solutions proposed by recent technologies for exploring genome and transcriptome alterations by mass and single-cell sequencing.

Single-cell experiments can be considered as thousands of separate experiments, so it is essential to apply the right quality control (QC) metrics to decide which individual data sets are valid [56]. For example, in a droplet-based experiment the QC can effectively determine, by applying a number of different parameters, which droplets are failed and exclude these data from further analysis [57,58]. An important QC metric to evaluate is the number of transcripts per cell, or the percentage of transcripts per cell that align with the reference genome and establish a cutoff to identify outliers. These cutoffs must be defined by the user for each experiment, for instance: cells with a few dozen transcripts and/or with several thousand; otherwise, they can be automatically defined by a software as cells with a summary of transcripts greater than two SD from the mean value. Excessive numbers of uniquely barcoded transcripts may result from duplicates (i.e., two or more cells suspended in a drop), whereas a small number of transcripts is an indicator of poor capture quality. Additional QC metrics related to the diversity of the tissue to be analyzed must then be applied [59,60]. For example, in an experiment to study circulating tumor cells, the number of tumor cells will be very low compared with normal blood cells and transcript counts will need to be adjusted; in fact, normal and generally quiescent blood cells have relatively low amounts of RNA compared with active tumor cells.

A common QC metric is the number of mitochondrial gene transcripts: excessive numbers of mitochondrial transcripts indicate cellular stress. In normal tissues, cells with excessive mitochondrial gene expression are not included in the analysis [61]. However, this parameter is highly dependent on the tissue and the purpose of the investigation [62]. Mitochondrial mRNA percentages should be assessed in a tissue-dependent approach.

An important point in analyzing single-cell data is normalization to eliminate batch effects if multiple sequencing runs are to be compared. These batch effects can be caused by a non-avoidable number of technical variations given by different experimental sessions (e.g., RNA isolation method, sequencing depth, etc.). In addition, for bulk RNA sequencing, data normalization involves comparing multiple batches of biological material; however, in sequencing individual cells that are not all the same type, normalization parameters are required to maintain cell-to-cell variability. A common way to normalize sequencing data is based on comparison with housekeeping genes [62,63]. Based on the characteristics of the biological sample, a selected housekeeping gene is chosen for normalization. Assuming that this gene is expressed at the same level in all cells, data are scaled to make the expression level of the housekeeping gene equal in all cells.

The next analytic step is to use a clustering algorithm to determine which cells are closely related. The most widely used is the principal component analysis (PCA) [60], which uses a relatively simple linear dimensionality reduction algorithm; the latter can predict the relatedness of cells in this case based solely on differential gene expression. Due to the highly dimensional nature of scRNA-seq data, several reduction methods are required, including nonlinear methods such as the t-Distributed stochastic neighbor embedding (t-SNE) and the uniform manifold approximation and projection (UMAP) techniques. The t-SNE is a common data visualization approach [64–66] that uses a machine learning algorithm to reduce size and is suitable for embedding high-density data into a two- or three-dimensional setting for visualization. For example, if cell diversity was found to be well represented with some PCs, t-SNE will plot the cells on a two-dimensional graph in a way that preserves the relationship between cells; as a consequence, cells that are close on a multi-dimensional graph remain close together on a two-dimensional graph. UMAP is a dimension reduction technique that can be used not only for visualization but also for general nonlinear dimension reduction [67]. Sensitivity studies on these methods determined that t-SNE gave the best overall performance with the highest accuracy. On the other hand, UMAP showed the highest stability and moderate accuracy while well retaining original cohesion and separation of cell populations [68].

3. Recent Results on Precision Medicine Applied to Colorectal Carcinoma

Intratumoral heterogeneity is a crucial factor in tumor biology, response to therapies and patient survival [69,70]. Due to the need to characterize the phenotypes and interactions of the tumoral cell subtypes, to date molecular profiling studies have adopted a bulk approach by not identifying the signatures of distinct cell populations.

As single-cell sequencing technologies ensure a complete, unbiased analysis of cellular diversity within tumor masses, they can be used to explore the measurement somatic mutation rates, the clonal evolution of cell tumor lineages, and gain insights into chemotherapeutic drug response [71,72]. Whole genomic and transcriptomic profiling of a tumor sample shows us only average measures of cellular characteristics, thus concealing critical aspects of tumor heterogeneity.

Currently, several studies on single cells genomics and transcriptomics analysis [6,73,74] have increased existing molecular classifications of CRC by detecting new distinct subclones within a single phenotype, previously identified through standard transcriptomics [31,75].

Dai et al. generated a molecular census of tumor tissue cell types of a single CRC patient alongside with a clustering analysis to define gene expression at single-cell level. A total of 2824 cells were identified and classified into five distinct cell clusters. Each cluster was characterized by different cell markers: cluster 2 prevalently contained genes related to the major histocompatibility complex, while the remaining 4 possessed cell markers related to themselves. Gene Ontology term analysis demonstrates that cluster 1 genes were responsible for biological processes including ATP synthesis, cellular respiration, and energy derivation. Cluster 3 and 4 genes mainly supported cells by providing energy, generating extracellular matrix. Cluster 2 and 5 genes highlighted immunity functions including immune response, regulation of lymphocyte, leukocyte, and T-cell activation. Although the results of Dai et al. were obtained by a single CRC patient, they help us understand how different activated and quiescent, abnormal cellular subpopulations contribute to the initiation, maintenance, and progression of CRC disease [75]. These data could represent an interactive map of genetic interaction and might be used to identify targets to develop new therapeutic options for CRC.

Li et al. performed an scRNA-seq analysis on 11 primary CRCs and matched normal mucosa to their microenvironments. They developed a method for single-cell transcriptome analysis defined reference component analysis (RCA) based on an algorithm that improves clustering accuracy.

Seven major cell types both in normal mucosa and CRC were isolated as well as epithelial cells, fibroblasts, endothelial cells, B cells, T cells, mast cells and myeloid cells. By using RCA, nine epithelial clusters and seven epithelial cell subtypes in human normal mucosa were isolated *de novo*. These reference data allow to identify a strong enrichment of stem/TA-like cells. Two distinct types of cancer-associated fibroblasts (CAFs), and epithelial–mesenchymal transition-related genes were found to be upregulated in the tumoral CAF subpopulation. CRC defined as single type in bulk transcriptomics, might be divided into subgroups with different survival probability rates by using single-cell signatures [75].

A recent study characterized the individual cell response of CRC cell lines to genotoxic 5-fluorouracil (5FU)-induced DNA damage using a scRNA-seq approach. After 5FU treatment, the apparently single population CRC cells assume three distinctive transcriptome profiles, corresponding to diversified cell-fate responses: apoptosis, cell-cycle checkpoint, and stress resistance. Based on the group-specific expression gene patterns mediating DNA damage responses, it can be inferred how individual cells shape their transcriptome in response to DNA damage involving recurrence and chemoresistance. This might represent one of the most important challenges in current cancer treatment [76]. The identification of cell-fate-specific transcriptome patterns in in vitro experiments should promote future studies on human CRC to explore heterogeneous cancer cell responses to genotoxic chemotherapy, such as fractional killing and chemoresistant tumor recurrence.

Metastasis is a complex biological process in which tumor cells move from the primary organ site and spread to distant organs through blood circulation [77]. Various models of metastasis have been proposed: late spread, early spread, and self-seeding. In the first one, tumor cells evolve over an extended stage at the primary site and then acquire specific mutations that allow them to spread. In contrast, in the second one, cancer cells spread early, and thus primary and metastatic tumors evolve in parallel [78]. Finally, based on the self-seeding hypothesis, tumor cells spread from the primary tumor establishing distant metastatic sites and then bidirectionally return to the primary site to promote growth [79].

A general difficulty in understanding metastatic lineages depends on the large intratumor heterogeneity at primary and metastatic sites. Leung et al. developed a highly multiplexed single-cell DNA sequencing approach to dissect the clonal evolution during the metastatic process. They studied two CRC patients with matching liver metastases. They observed monoclonal seeding in the first patient: a single clone acquired a large number of mutations before migrating to the liver to establish the second tumor site. In the second patient, they observed polyclonal seeding: two independent clones seeded metastases to the liver after migrating from the primary tumor lineage at different time points. Single-cell data also revealed a striking independent tumor lineage that did not metastasize, and early progenitor clones with the *"first hit"* mutation in APC that subsequently gave rise to both the primary and metastatic tumors. Data from this study revealed a late-dissemination model of metastasis in both CRC patients and provided unprecedented insight into metastasis at single-cell genomic resolution [80]. Actually, despite the small number of CRC patients observed and the fact that only the liver metastatic site was examined, Leung's study represents a preliminary confirmation that late-dissemination models of metastasis can occur in CRC but should not be contemplated as a common model for all CRC patients.

Tang et al. characterized the evolutionary pattern of metastatic CRC (mCRC) by analyzing bulk and single-cell whole-exome sequencing (scWES) data of primary and metastatic tumors from seven CRC patients. They proved that genomic profile could be better explained by using scWES than through bulk sequencing. Rare mutations highlighted by scWES were undetectable in bulk data. Several subclones have been identified in both primary and metastatic tumor cells in MSI CRC patients. Although the individual cells of each subclone share a substantial number of mutations, few subclone-specific single nucleotide variants (SNVs) could characterize different cell clones with low mutation frequencies in the entire population of tumor cells.

In MSS CRC patients, tumor cells were divided into two major cell populations from primary and metastatic lesions, that shared most SNVs and involved genes associated with CRC progression, such as TP53 and APC. Primary tumor cell populations were rich in AXIN3 and RASGRF1 genes mutation, known to be associated with tumor proliferation and invasion. In addition, 24 non-synonymous SNVs specific to metastatic cells in DNAH3, TBC1D4, CMYA5, MYO18A, PLEKHA7, and SLC19A3 genes have been identified, validating their functions in cell migration capacity [81].

Another comparison of scWES versus bulk whole-exome sequencing (bulk WES) on two CRC patients with tumor and adenomatous polyps, showed that both had monoclonal origin and shared partial mutations in the same signaling pathways; however, each showed a specific spectrum of heterogeneous somatic mutations. Adenoma and cancer further developed intratumor heterogeneity accumulating non-random somatic mutations specifically in GPCR, PI3K-Akt and FGFR signaling pathways. New driver mutations were identified that developed during the evolution of both adenoma and cancer: on one hand OR1B1 (GPCR signaling pathway) was related to adenoma evolution; on the other hand, LAMA1 (PI3K-Akt signaling pathway) and ADCY3 (FGFR signaling pathway) had a role in CRC evolution. ScWES shows causality of mutations in certain pathways that would not be detected by bulk tumor sequencing. Furthermore, it can potentially establish whether specific mutations are mutually exclusive or occur sequentially in the same subclone of cells [82].

To examine the genome, transcriptome, and methylome within CRC primary tumors and metastases, Bian et al. used a single-cell triple homology sequencing (scTrio-seq) technique [83]. The scTrio-seq technique can assess somatic copy number alterations (SCNA), as well as DNA methylation and transcriptome information simultaneously from the same single cell [84]. The authors performed a multiregional sampling and generated scTrio-seq profiles for 12 CRC patients with stage III or IV cancer. The majority of tumor cells from six of the patients analyzed were assigned to the group with abnormal activation of WNT/β-catenin and MYC signaling pathways, frequent somatic copy number alterations (SCNAs), and no hypermutation. In the 10 patients with DNA methylation data were relatively consistent within a single genetic line, single-cell SCNA profiling identified significant focal SCNAs and likely target genes. Differences in methylation profiles between primary and metastatic sites could be primarily due to differences in sub-lineage composition. No results from *de novo* methylation or demethylation during metastasis were observed. As well as providing important information about the molecular alterations that occur during CRC progression and metastasis, multicellular sequencing showed that DNA methylation levels are consistent within lineages but can differ substantially between clones [83].

To summarize, shedding light on the main mechanisms behind the development of metastasis based on the analysis of gene expression patterns at single-cell resolution should lead to tailoring individualized cancer treatment.

To this end, the study of CRC heterogeneity through identification of tumor cells subpopulations and analysis of their features by single-cell omics technologies is crucial for the comprehension of the role of these cells and might lead to identify potential new targets for clinical treatment.

Table 1 provides a summary of the most recent advances of the application of both single-cell sequencing and editing technologies into precision medicine applied to CRC patients.

Application of omics technologies on other types of cancers is opening a way to verify results also in diagnosis and treatment of other tumoral diseases, including CRC; this is the case of breast cancer (BC) thoroughly studied through single-cell omics technologies. For instance, Pinkney et al. adopted scRNA-seq to analyze the heterogeneity of lncRNA expression in vivo using Triple Negative BC (TNBC) xenografts; at the same time, they tried to assess whether lncRNA expression is sufficient to define cellular subpopulations. These authors observed that even if most lncRNAs are detectable at low levels in TNBC xenografts, a subpopulation of cells could not be defined. They showed highly heterogeneous expression patterns including global expression and subpopulation-specific expression; in addition, a hybrid pattern of lncRNAs was expressed in several but not all subpopulations [85].

LncRNAs have been progressively identified as the main group of oncology targets acting as drivers in cancer, and are also being studied as clinical biomarkers [86]. LncRNAs link with biological molecules, as well as with DNA, mRNAs, miRNAs and proteins, modulating epigenetic, transcriptional, post-transcriptional, translational and post-translational events in gene expression [87,88]. LncRNAs have been observed to be of interest in cancer, but little is known about their expression in cell subpopulations. Further investigation may determine whether expression of specific lncRNAs contribute to specific cell populations features; they might have a role also in invasion and/or proliferation, considering that lncRNAs have been described as drivers of these processes [89]. Therefore, the spatial distribution of lncRNAs within a patient's cancer tissues might identify the potential of subclone-specific lncRNAs as new therapeutic targets and/or biomarkers.

Zhang et al. performed a single-cell RNA- and ATAC-sequencing to examine the immune cell dynamics in advanced TNBC patients treated with paclitaxel or paclitaxel plus atezolizumab (anti-PD-L1). High levels of baseline CXCL13+ T cells linked to macrophage proinflammatory features might predict responses to a drug combination. In patients responsive to drug combination, an increase of lymphoid tissue inducer cells, follicular B cells, CXCL13+ T cells, and type 1 dendritic cells was detected. The latter decreased after paclitaxel monotherapy [90]. These data suggest the role of CXCL13+ T cells in the responses to anti-PD-L1 therapies.

Immune checkpoint blockade (ICB) targeting PD-1/PD-L1 signaling axis and its use has achieved significant responses in cancer patients, although the mechanisms underlying ICB resistance have not been fully understood [91,92]. Thus, the advances in single-cell technologies enable to characterize the basic properties of tumor-infiltrating immune cells to determine their role in immune responses, antitumor immunity, and immunotherapies.

Table 1. Summary of advances of single-cell sequencing and editing technologies into precision medicine in the colorectal cancer.

Sample Type	Technology	Findings	Implications	Ref.
1 patients 2824 sc	scRNA-seq	-5 distinct cell subsets were identified consisting of: immune cells, related to the major histocompatibility complex genes, related to genes serving to stabilize the cell, energy transportation and cell regulation, TSPAN6, PFDN4, and TIMM13, majored in breakdown of extracellular matrix and tissues remodeling, and genes involved in cancer, WFDC2 -cluster 1 and 3 revealed biological processes genes, including ATP synthesis, cellular respiration, oxidative phosphorylation, and mitochondrion organization -cluster 2 and 5 revealed biological process genes, consisting of activation, positive regulation, response to stress, cellular response, and cell adhesion -cluster 4 revealed biological processes responsible for extracellular matrix organization, response to stress, locomotion, cell migration, and cell motility	-provides insight into the heterogeneity of CRC and which genes within each cluster serve different functions	[93]
11 patients 7 cell lines CRC 590 patient-derived sc 561 cell line-derived sc	scRNA-seq, reference component analysis algorithm	-scRNA-seq generated further sub-classification of CRC subtypes found by bulk RNA-seq with prognostic significance based on their single-cell signatures	-scRNA-seq could enable clinically relevant patient stratification	[75]
3 cell lines CRC	scRNA-seq	-transcriptomic characterization of CRC cell lines response to 5-fluorouracil (5FU)-induced DNA damage -three distinct transcriptome phenotypes were assumed by CRC cells, with different cell-fate responses: apoptosis, cell-cycle checkpoint, and stress resistance	-understanding of the heterogeneous DNA damage responses involved in fractional killing and chemoresistance	[76]
2 patients 360 sc and bulk primary tumor and liver metastasis	scNGS, bulk WES	-the single-cell and bulk analyses were highly concordant -monoclonal and polyclonal seeding were found -rare cell subpopulations were associated with progression and metastasis -a late-dissemination model was highly concordant between primary tumor and liver metastasis samples	-the late-dissemination model suggests that early surgical intervention could prevent metastasis	[80]
7 patients 321 sc and bulk primary tumor and liver metastasis	scWES, bulk WES	-low genomic divergence between paired primary and metastatic cancers were found in bulk data -scWES data defined two separate cell populations, indicative of the diverse evolutionary trajectories between primary and metastatic tumor cells. -rare mutations were identified using single-cell technology that were overlooked in bulk data	-validation of functions of different metastatic subclone-specific-mutated genes in cell migration	[81]
2 patients 96 sc (adenomatous polyp and CRC)	scWES, bulk WES	-adenoma and cancer have monoclonal origin with subsequent subclonal evolution -adenoma and cancer showed a specific spectrum of heterogeneous somatic mutations -novel driver mutations that developed during adenoma and cancer evolution, in OR1B1 (GPCR signaling pathway) for adenoma evolution; LAMA1 (PI3K-Akt signaling pathway) and ADCY3 (FGFR signaling pathway) for CRC evolution	-scWES provides evidence for the importance of mutations in certain pathways that would not be so apparent from bulk sequencing of tumors	[82]
12 patients 1900 sc and bulk multi-regional	scTrio-seq, bulk multi-regional WGS	-cancer cells were classified into several genetic subclones -primary tumor showed higher subclonality than metastatic tumour -DNA methylation profiles were stable within a single genetic lineage	-single-cell multiomics sequencing can trace epigenomic and transcriptomic dynamics during progression and metastasis	[83]

Table 1. *Cont.*

Sample Type	Technology	Findings	Implications	Ref.
2 patients CRC clonal tumor organoids	3D Live-Seq (a protocol that integrates live-cell imaging of tumor organoid outgrowth and WGS of each imaged cell to reconstruct evolving tumor cell karyotypes across consecutive cell generations)	-reveals the genomic consequences of CIN across consecutive cell generations -single-cell sequencing data displayed several de novo CNAs across three lineages -mis-segregation of chromosome 7 displays the highlighted branch within the mitotic tree	-mapping the temporal dynamics and patterns of karyotype diversification in cancer enables reconstructions of evolutionary paths to malignant fitness	[94]
Cell lines CRC tumor, stroma, adjacent normal, lung metastasis	quantitative micro-engraving	-single cells exhibit a range of secretory phenotypes for CXCL1, CXCL5, and CXCL8 -secretions of ELR+ CXC chemokines were found from thousands of single CRC and stromal cells -CRC and stromal cells exhibit polyfunctional heterogeneity in the combinations and magnitudes of secretions for these chemokines -discordances exist between secretory states measured and gene expression for these chemokines among single cells	-these measures suggest that secretory states among tumor cells are complex and can dynamically evolve -heterogeneous release of these chemokines by individual cells promotes a robust signaling network within the tumor microenvironment	[95]
14 patients 336 cells each phenotypic population	scPCR gene-expression analysis	-CRC tissues contain distinct cell populations whose transcriptional identities mirror those of the different cellular lineages in healthy colon -perturbations in gene expression programs linked to multi-lineage differentiation strongly associate with patient survival -development of two-gene classifier systems (KRT20 vs. CA1, MS4A12, CD177, SLC26A3) that predict clinical outcomes with hazard-ratios superior to pathological grade	-development of a simple and quantitative nature two-gene scoring system	[96]
2 patients 88 sc rectal cancer	WES, scWGS multi-region	-genomic heterogeneity was observed between the two patients, and the degree of ITH increased when analyzed at single-cell level -SCNAs were early events in cancer development -single-cell sequencing revealed mutations and SCNAs which were hidden in bulk sequencing	-each tumor possesses its own architecture, which may result in different diagnosis, -prognosis, and drug responses	[97]
2 patients 47 sc cancer stem and differentiated tumor	scWGS	-CD45− EpCAM^{high} CD44+ CSCs and CD45− EpCAM^{high} CD44− differentiated tumor cells had similar SCNA profiles -the similarity of ubiquitous SCNAs between the CSCs and DTCs might have arisen from lineage differentiation	-the possibility of a monoclonal CSC phenotype is supported	[98]
3 patients organoid from multiple sc CRC and normal mucosa	scWGS	-significant intra-tumor clonal heterogeneity with specific mutational signatures were identified organoids treated with chemotherapeutic and targeted agents, even derived from the same patient, exhibited differential responses independent of their mutational signatures	-substantial increases in somatic mutation rate compared to normal colorectal cells -genetic diversification of each cancer is accompanied by pervasive, stable, and inherited differences in biological states of individual cancer cells	[6]

scRNA-seq: single-cell RNA-seq, sc: single-cell, CRC: colorectal carcinoma, scNGS: single-cell next generation sequencing, scWES: single-cell whole-exome sequencing, WGS: whole genome sequencing, SCNAs: somatic copy number alterations, TME: tumor microenvironment, S-TAM: small tumor-associated macrophages, L-TAM: large tumor-associated macrophages, CCDGs: cell cluster deregulated genes, CIN: chromosomal instability, CAN: copy-number alterations, ITH: intratumor heterogeneity.

4. Conclusions

4.1. Future Perspectives in Methodologies

In-depth knowledge of the cells of interest is crucial to properly manage genomic data and make decisions of clinical impact based on standardized measurements and accurate and reproducible quality controls. The use of scRNA-seq provides one of the most innovative methods for addressing biological and medical questions concerning the underlying processes of various developmental, physiological, and disease systems. However, new programs and implementations of scRNA-seq methodologies have been started in recent years but further advances in both technology and specific approaches to use them are certainly warranted.

The deployment of a number of processes will make it possible to extend the analysis of scRNA-seq studies not only on fresh material, but also on cryopreserved and fixed tissue samples aiming at introducing this technique into the clinical practice. Volume reduction and diffusion of techniques based primarily on microfluidics platforms should reduce costs at the same time leading to a standardized and simplified use of different devices.

However, one of the current challenges is the creation of standardized collections and data catalogues from single cells due to the fact that the number of samples used so far in studies is small. Such analysis, in fact, requires a minimum/sufficient number of cells to ensure that all cell types are represented. Only a bioinformatician with experience in single-cell sequencing will be able to generate analyses that can be used to make meaningful biological inferences by choosing appropriate cutoffs for applied algorithms and avoiding misleading results. Currently, there are limited standardization protocols and guidelines on standards (i.e., quality control, removal of technical artifacts, etc.).

Furthermore, development of single-cell gene expression maps for all tissues will be necessary, as it occurred in bulk transcriptomics evolution. Many studies, in fact, will benefit from these easily accessible archives that reduce the costs of comparison and replication in normal tissues; at the same time, significant advances in bioinformatics and computational methods thanks to data sharing are expected.

Thus, the new challenge will be represented by the use of a true inter-omic and multidisciplinary approach that will lead to a comprehensive examination of individual cells; this will be achieved by characterizing the genome, epigenome, proteome and metabolome while simultaneously examining the tumor microenvironment, its immunological characteristics and the impact of pharmacogenomics; in addition, a clear picture of tumor development will be given, together with cancer evolution and interactions. It will be crucial to address genetic changes in the early stages of tumorigenesis deployment and how transcriptional subpopulations evolve into malignancy in later stages of tumor progression.

The robustness of NGS systems in exploring heterogeneity, at genome and transcriptome scale, will validate ITH variability and might determine the discovery of novel targeted drugs; predictive biomarkers for individualized drug-oriented therapies might also be developed. Pharmacogenomic profiling might predict response to chemotherapy by correlating it with immune cell regulatory values that affect CRC survival mechanisms. Future CRC studies employing comparison of primary, metastatic tumor ITH and liquid biopsies might offer elucidating suggestions on the origins and evolution of genomic subclones responsible for drug resistance and recurrence.

4.2. Clinical Implications of Intra-Tumoral Heterogeneity in CRC

CRC is extensively marked by phenomena of inter- and intra-tumor heterogeneity, spatial and temporal differences regarding phenotypic and genotypic aspects, influencing recurrence and therapeutic response and having a strong poor impact on CRC patient's outcome.

Until now, genomic and transcriptome analyses on bulk tumor cell populations have helped to explain tumor heterogeneity and also allowed to classify them into subgroups with distinct molecular, morphological, and clinical features [99]. The application of techniques capable of examining molecular aberrations at the single-cell level within a complex tumor population should refine the existing CRC classification system. In addition, scRNA-seq could identify predictive markers for CRC prognosis.

Metastatic progression is linked to the majority of CRC-related deaths [100]. In patients at stage I, the five-year survival rate is 90%, but a drastic reduction of slightly more than 10% is observed when cancer patients reach stage IV [101]. Approximately 20% of CRC patients already have metastases at diagnosis, and they are generally incurable [100]. Although anti-EGFR therapies are available for RAS wild-type CRC patients, and anti-VEGF, anti-VEGFR, recombinant fusion protein and multi-kinase inhibitor were applied in CRC patients with RAS mutation [102], unresponsiveness was seen in CRC patients with BRAF and PIK3CA mutations [103]. Undoubtedly, drug development and techniques to be used in identifying the complex heterogeneity of mCRC represent an unmet clinical need. Single-cell omics represent an important tool to identify therapeutic targets for personalized cancer medicine compared with bulk transcriptomics. In addition, single-cell resolution molecular aberrations could shed light on the mechanisms underlying metastasis development [104–106]. Finally, the ability to estimate presence of rare malignant chemical-resistant carcinoma cells in removed tumors will be increased to guide treatment decisions; at the same time exploration of immune cell responses and environmental influences will provide molecular data to give support during the diagnostic process as well as in disease progression, and treatment course.

Author Contributions: Conceptualization, M.R.D.M., A.A. and A.M.S.; literature survey and database curation, M.R.M. and C.A.; writing—original draft preparation, M.R.D.M. and A.A.; writing—review and editing the full content of the manuscript, C.C. and A.P.; funding acquisition, M.R.D.M., A.P. and A.M.S.; supervision, P.C.-R. and A.P. All authors have read and agreed to the published version of the manuscript.

Funding: This work was partly supported by grants from Fondazione Banco di Sardegna, Italy, and Fondo di Beneficenza, Intesa Sanpaolo S.p.A. (Milano, Italy) Grant Number: B/2020/0094.

Institutional Review Board Statement: Not applicable.

Informed Consent Statement: Not applicable.

Data Availability Statement: Not applicable.

Conflicts of Interest: The authors declare no conflict of interest.

References

1. Shendure, J.; Balasubramanian, S.; Church, G.M.; Gilbert, W.; Rogers, J.; Schloss, J.A.; Waterston, R.H. DNA sequencing at 40: Past, present and future. *Nature* **2017**, *550*, 345–353. [CrossRef]
2. Cao, J.; Spielmann, M.; Qiu, X.; Huang, X.; Ibrahim, D.M.; Hill, A.J.; Zhang, F.; Mundlos, S.; Christiansen, L.; Steemers, F.J.; et al. The single-cell transcriptional landscape of mammalian organogenesis. *Nature* **2019**, *566*, 496–502. [CrossRef]
3. Knott, G.J.; Doudna, J.A. CRISPR-Cas guides the future of genetic engineering. *Science* **2018**, *361*, 866–869. [CrossRef]
4. Porteus, M.H. A New Class of Medicines through DNA Editing. *N. Engl. J. Med.* **2019**, *380*, 947–959. [CrossRef] [PubMed]
5. Grünewald, J.; Zhou, R.; Garcia, S.P.; Iyer, S.; Lareau, C.A.; Aryee, M.J.; Joung, J.K. Transcriptome-wide off-target RNA editing induced by CRISPR-guided DNA base editors. *Nature* **2019**, *569*, 433–437. [CrossRef] [PubMed]
6. Roerink, S.F.; Sasaki, N.; Lee-Six, H.; Young, M.D.; Alexandrov, L.B.; Behjati, S.; Mitchell, T.J.; Grossmann, S.; Lightfoot, H.; Egan, D.A.; et al. Intra-tumour diversification in colorectal cancer at the single-cell level. *Nature* **2018**, *556*, 457–462. [CrossRef]
7. Khan, K.H.; Cunningham, D.; Werner, B.; Vlachogiannis, G.; Spiteri, I.; Heide, T.; Mateos, J.F.; Vatsiou, A.; Lampis, A.; Damavandi, M.D.; et al. Longitudinal liquid biopsy and mathematical modeling of clonal evolution forecast time to treatment failure in the prospect-c phase ii colorectal cancer clinical trial. *Cancer Discov.* **2018**, *8*, 1270–1285. [CrossRef] [PubMed]
8. Mamlouk, S.; Childs, L.H.; Aust, D.; Heim, D.; Melching, F.; Oliveira, C.; Wolf, T.; Durek, P.; Schumacher, D.; Bläker, H.; et al. DNA copy number changes define spatial patterns of heterogeneity in colorectal cancer. *Nat. Commun.* **2017**, *8*, 14093. [CrossRef] [PubMed]
9. Strickler, J.H.; Loree, J.M.; Ahronian, L.G.; Parikh, A.R.; Niedzwiecki, D.; Pereira, A.A.L.; McKinney, M.; Michael Korn, W.; Atreya, C.E.; Banks, K.C.; et al. Genomic landscape of cell-free DNA in patients with colorectal cancer. *Cancer Discov.* **2018**, *8*, 164–173. [CrossRef]
10. Jones, P.A.; Baylin, S.B. The Epigenomics of Cancer. *Cell* **2007**, *128*, 683–692. [CrossRef]
11. You, J.S.; Jones, P.A. Cancer Genetics and Epigenetics: Two Sides of the Same Coin? *Cancer Cell* **2012**, *22*, 9–20. [CrossRef] [PubMed]

12. Rodon, J.; Soria, J.C.; Berger, R.; Miller, W.H.; Rubin, E.; Kugel, A.; Tsimberidou, A.; Saintigny, P.; Ackerstein, A.; Braña, I.; et al. Genomic and transcriptomic profiling expands precision cancer medicine: The WINTHER trial. *Nat. Med.* **2019**, *25*, 751–758. [CrossRef] [PubMed]
13. Rothwell, D.G.; Ayub, M.; Cook, N.; Thistlethwaite, F.; Carter, L.; Dean, E.; Smith, N.; Villa, S.; Dransfield, J.; Clipson, A.; et al. Utility of ctDNA to support patient selection for early phase clinical trials: The TARGET study. *Nat. Med.* **2019**, *25*, 738–743. [CrossRef] [PubMed]
14. Sicklick, J.K.; Kato, S.; Okamura, R.; Schwaederle, M.; Hahn, M.E.; Williams, C.B.; De, P.; Krie, A.; Piccioni, D.E.; Miller, V.A.; et al. Molecular profiling of cancer patients enables personalized combination therapy: The I-PREDICT study. *Nat. Med.* **2019**, *25*, 744–750. [CrossRef] [PubMed]
15. Jamal-Hanjani, M.; Wilson, G.A.; McGranahan, N.; Birkbak, N.J.; Watkins, T.B.K.; Veeriah, S.; Shafi, S.; Johnson, D.H.; Mitter, R.; Rosenthal, R.; et al. Tracking the Evolution of Non–Small-Cell Lung Cancer. *N. Engl. J. Med.* **2017**, *376*, 2109–2121. [CrossRef] [PubMed]
16. Angius, A.; Scanu, A.M.; Arru, C.; Muroni, M.R.; Rallo, V.; Deiana, G.; Ninniri, M.C.; Carru, C.; Porcu, A.; Pira, G.; et al. Portrait of cancer stem cells on colorectal cancer: Molecular biomarkers, signaling pathways and mirnaome. *Int. J. Mol. Sci.* **2021**, *22*, 1603. [CrossRef]
17. McGranahan, N.; Swanton, C. Biological and therapeutic impact of intratumor heterogeneity in cancer evolution. *Cancer Cell* **2015**, *27*, 15–26. [CrossRef] [PubMed]
18. Bedard, P.L.; Hansen, A.R.; Ratain, M.J.; Siu, L.L. Tumour heterogeneity in the clinic. *Nature* **2013**, *501*, 355–364. [CrossRef]
19. Cusnir, M.; Cavalcante, L. Inter-tumor heterogeneity. *Hum. Vaccines Immunother.* **2012**, *8*, 1143–1145. [CrossRef]
20. Navin, N.; Kendall, J.; Troge, J.; Andrews, P.; Rodgers, L.; McIndoo, J.; Cook, K.; Stepansky, A.; Levy, D.; Esposito, D.; et al. Tumour evolution inferred by single-cell sequencing. *Nature* **2011**, *472*, 90–94. [CrossRef]
21. Gerlinger, M.; Rowan, A.J.; Horswell, S.; Larkin, J.; Endesfelder, D.; Gronroos, E.; Martinez, P.; Matthews, N.; Stewart, A.; Tarpey, P.; et al. Intratumor Heterogeneity and Branched Evolution Revealed by Multiregion Sequencing. *N. Engl. J. Med.* **2012**, *366*, 883–892. [CrossRef] [PubMed]
22. Driessens, G.; Beck, B.; Caauwe, A.; Simons, B.D.; Blanpain, C. Defining the mode of tumour growth by clonal analysis. *Nature* **2012**, *488*, 527–530. [CrossRef] [PubMed]
23. Muzny, D.M.; Bainbridge, M.N.; Chang, K.; Dinh, H.H.; Drummond, J.A.; Fowler, G.; Kovar, C.L.; Lewis, L.R.; Morgan, M.B.; Newsham, I.F.; et al. Comprehensive molecular characterization of human colon and rectal cancer. *Nature* **2012**, *487*, 330–337. [CrossRef]
24. Guinney, J.; Dienstmann, R.; Wang, X.; De Reyniès, A.; Schlicker, A.; Soneson, C.; Marisa, L.; Roepman, P.; Nyamundanda, G.; Angelino, P.; et al. The consensus molecular subtypes of colorectal cancer. *Nat. Med.* **2015**, *21*, 1350–1356. [CrossRef]
25. Hinoue, T.; Weisenberger, D.J.; Lange, C.P.E.; Shen, H.; Byun, H.M.; Van Den Berg, D.; Malik, S.; Pan, F.; Noushmehr, H.; Van Dijk, C.M.; et al. Genome-scale analysis of aberrant DNA methylation in colorectal cancer. *Genome Res.* **2012**, *22*, 271–282. [CrossRef]
26. Berman, B.P.; Weisenberger, D.J.; Aman, J.F.; Hinoue, T.; Ramjan, Z.; Liu, Y.; Noushmehr, H.; Lange, C.P.E.; Van Dijk, C.M.; Tollenaar, R.A.E.M.; et al. Regions of focal DNA hypermethylation and long-range hypomethylation in colorectal cancer coincide with nuclear laminag-associated domains. *Nat. Genet.* **2012**, *44*, 40–46. [CrossRef] [PubMed]
27. Angius, A.; Pira, G.; Scanu, A.M.; Uva, P.; Sotgiu, G.; Saderi, L.; Manca, A.; Serra, C.; Uleri, E.; Piu, C.; et al. Microrna-425-5p expression affects BRAF/RAS/MAPK pathways in colorectal cancers. *Int. J. Med. Sci.* **2019**, *16*, 1480–1491. [CrossRef]
28. Angius, A.; Uva, P.; Pira, G.; Muroni, M.R.; Sotgiu, G.; Saderi, L.; Uleri, E.; Caocci, M.; Ibba, G.; Cesaraccio, M.R.; et al. Integrated Analysis of miRNA and mRNA endorses a twenty miRNAs signature for colorectal carcinoma. *Int. J. Mol. Sci.* **2019**, *20*, 4067. [CrossRef]
29. Pira, G.; Uva, P.; Scanu, A.M.; Rocca, P.C.; Murgia, L.; Uleri, E.; Piu, C.; Porcu, A.; Carru, C.; Manca, A.; et al. Landscape of transcriptome variations uncovering known and novel driver events in colorectal carcinoma. *Sci. Rep.* **2020**, *10*, 432. [CrossRef]
30. Haendel, M.A.; Chute, C.G.; Robinson, P.N. Classification, Ontology, and Precision Medicine. *N. Engl. J. Med.* **2018**, *379*, 1452–1462. [CrossRef]
31. Suvà, M.L.; Tirosh, I. Single-Cell RNA Sequencing in Cancer: Lessons Learned and Emerging Challenges. *Mol. Cell* **2019**, *75*, 7–12. [CrossRef] [PubMed]
32. Bray, F.; Ferlay, J.; Soerjomataram, I.; Siegel, R.L.; Torre, L.A.; Jemal, A. Global cancer statistics 2018: GLOBOCAN estimates of incidence and mortality worldwide for 36 cancers in 185 countries. *CA Cancer J. Clin.* **2018**, *68*, 394–424. [CrossRef] [PubMed]
33. Eirew, P.; Steif, A.; Khattra, J.; Ha, G.; Yap, D.; Farahani, H.; Gelmon, K.; Chia, S.; Mar, C.; Wan, A.; et al. Dynamics of genomic clones in breast cancer patient xenografts at single-cell resolution. *Nature* **2014**, *518*, 422–426. [CrossRef] [PubMed]
34. Uchi, R.; Takahashi, Y.; Niida, A.; Shimamura, T.; Hirata, H.; Sugimachi, K.; Sawada, G.; Iwaya, T.; Kurashige, J.; Shinden, Y.; et al. Integrated Multiregional Analysis Proposing a New Model of Colorectal Cancer Evolution. *PLoS Genet.* **2016**, *12*, e1005778. [CrossRef]
35. Cristiano, S.; Leal, A.; Phallen, J.; Fiksel, J.; Adleff, V.; Bruhm, D.C.; Jensen, S.Ø.; Medina, J.E.; Hruban, C.; White, J.R.; et al. Genome-wide cell-free DNA fragmentation in patients with cancer. *Nature* **2019**, *570*, 385–389. [CrossRef]
36. Ziogas, D.E.; Kyrochristos, I.D.; Roukos, D.H. Discovering novel valid biomarkers and drugs in patient-centric genomic trials: The new epoch of precision surgical oncology. *Drug Discov. Today* **2018**, *23*, 1848–1872. [CrossRef]

37. Liu, C.; Banister, C.E.; Weige, C.C.; Altomare, D.; Richardson, J.H.; Contreras, C.M.; Buckhaults, P.J. PRDM1 silences stem cell-related genes and inhibits proliferation of human colon tumor organoids. *Proc. Natl. Acad. Sci. USA* **2018**, *115*, E5066–E5075. [CrossRef]
38. Fellmann, C.; Gowen, B.G.; Lin, P.C.; Doudna, J.A.; Corn, J.E. Cornerstones of CRISPR-Cas in drug discovery and therapy. *Nat. Rev. Drug Discov.* **2017**, *16*, 89–100. [CrossRef]
39. Gasperini, M.; Hill, A.J.; McFaline-Figueroa, J.L.; Martin, B.; Kim, S.; Zhang, M.D.; Jackson, D.; Leith, A.; Schreiber, J.; Noble, W.S.; et al. A Genome-wide Framework for Mapping Gene Regulation via Cellular Genetic Screens. *Cell* **2019**, *176*, 377–390.e19. [CrossRef]
40. Macosko, E.Z.; Basu, A.; Satija, R.; Nemesh, J.; Shekhar, K.; Goldman, M.; Tirosh, I.; Bialas, A.R.; Kamitaki, N.; Martersteck, E.M.; et al. Highly parallel genome-wide expression profiling of individual cells using nanoliter droplets. *Cell* **2015**, *161*, 1202–1214. [CrossRef]
41. Kitzman, J.O. Haplotypes drop by drop. *Nat. Biotechnol.* **2016**, *34*, 296–298. [CrossRef] [PubMed]
42. Theilgaard-Mönch, K.; Cowland, J.; Borregaard, N. Profiling of gene expression in individual hematopoietic cells by global mRNA amplification and slot blot analysis. *J. Immunol. Methods* **2001**, *252*, 175–189. [CrossRef]
43. Nguyen, Q.H.; Lukowski, S.W.; Chiu, H.S.; Senabouth, A.; Bruxner, T.J.C.; Christ, A.N.; Palpant, N.J.; Powell, J.E. Single-cell RNA-seq of human induced pluripotent stem cells reveals cellular heterogeneity and cell state transitions between subpopulations. *Genome Res.* **2018**, *28*, 1053–1066. [CrossRef] [PubMed]
44. Nguyen, Q.H.; Pervolarakis, N.; Blake, K.; Ma, D.; Davis, R.T.; James, N.; Phung, A.T.; Willey, E.; Kumar, R.; Jabart, E.; et al. Profiling human breast epithelial cells using single cell RNA sequencing identifies cell diversity. *Nat. Commun.* **2018**, *9*, 1–12. [CrossRef]
45. Moon, H.S.; Je, K.; Min, J.W.; Park, D.; Han, K.Y.; Shin, S.H.; Park, W.Y.; Yoo, C.E.; Kim, S.H. Inertial-ordering-assisted droplet microfluidics for high-throughput single-cell RNA-sequencing. *Lab Chip* **2018**, *18*, 775–784. [CrossRef]
46. Goetz, J.J.; Trimarchi, J.M. Transcriptome sequencing of single cells with Smart-Seq. *Nat. Biotechnol.* **2012**, *30*, 763–765. [CrossRef]
47. Zhu, Y.Y.; Machleder, E.M.; Chenchik, A.; Li, R.; Siebert, P.D. Reverse transcriptase template switching: A SMART^{TM} approach for full-length cDNA library construction. *Biotechniques* **2001**, *30*, 892–897. [CrossRef]
48. Picelli, S.; Björklund, Å.K.; Faridani, O.R.; Sagasser, S.; Winberg, G.; Sandberg, R. Smart-seq2 for sensitive full-length transcriptome profiling in single cells. *Nat. Methods* **2013**, *10*, 1096–1098. [CrossRef]
49. Picelli, S.; Faridani, O.R.; Björklund, Å.K.; Winberg, G.; Sagasser, S.; Sandberg, R. Full-length RNA-seq from single cells using Smart-seq2. *Nat. Protoc.* **2014**, *9*, 171–181. [CrossRef]
50. Klein, A.M.; Mazutis, L.; Akartuna, I.; Tallapragada, N.; Veres, A.; Li, V.; Peshkin, L.; Weitz, D.A.; Kirschner, M.W. Droplet barcoding for single-cell transcriptomics applied to embryonic stem cells. *Cell* **2015**, *161*, 1187–1201. [CrossRef]
51. Hanson, W.M.; Chen, Z.; Jackson, L.K.; Attaf, M.; Sewell, A.K.; Heemstra, J.M.; Phillips, J.D. Reversible Oligonucleotide Chain Blocking Enables Bead Capture and Amplification of T-Cell Receptor α and β Chain mRNAs. *J. Am. Chem. Soc.* **2016**, *138*, 11073–11076. [CrossRef] [PubMed]
52. Liu, N.; Jiang, Y.; Xing, M.; Zhao, B.; Hou, J.; Lim, M.; Huang, J.; Luo, X.; Han, L. Digital Gene Expression Profiling Analysis of Aged Mice under Moxibustion Treatment. *Evid. Based Complement. Altern. Med.* **2018**, *2018*, 4767328. [CrossRef]
53. Islam, S.; Kjällquist, U.; Moliner, A.; Zajac, P.; Fan, J.B.; Lönnerberg, P.; Linnarsson, S. Characterization of the single-cell transcriptional landscape by highly multiplex RNA-seq. *Genome Res.* **2011**, *21*, 1160–1167. [CrossRef] [PubMed]
54. Islam, S.; Zeisel, A.; Joost, S.; La Manno, G.; Zajac, P.; Kasper, M.; Lönnerberg, P.; Linnarsson, S. Quantitative single-cell RNA-seq with unique molecular identifiers. *Nat. Methods* **2014**, *11*, 163–166. [CrossRef] [PubMed]
55. Lun, A.T.L.; McCarthy, D.J.; Marioni, J.C. A step-by-step workflow for low-level analysis of single-cell RNA-seq data with Bioconductor. *F1000Research* **2016**, *5*, 2122. [CrossRef]
56. Zhao, C.; Hu, S.; Huo, X.; Zhang, Y. Dr.seq2: A quality control and analysis pipeline for parallel single cell transcriptome and epigenome data. *PLoS ONE* **2017**, *12*, e0180583. [CrossRef]
57. Alles, J.; Karaiskos, N.; Praktiknjo, S.D.; Grosswendt, S.; Wahle, P.; Ruffault, P.-L.; Ayoub, S.; Schreyer, L.; Boltengagen, A.; Birchmeier, C.; et al. Cell fixation and preservation for droplet-based single-cell transcriptomics. *BMC Biol.* **2017**, *15*, 44. [CrossRef]
58. Ilicic, T.; Kim, J.K.; Kolodziejczyk, A.A.; Bagger, F.O.; McCarthy, D.J.; Marioni, J.C.; Teichmann, S.A. Classification of low quality cells from single-cell RNA-seq data. *Genome Biol.* **2016**, *17*, 29. [CrossRef]
59. Guo, M.; Wang, H.; Potter, S.S.; Whitsett, J.A.; Xu, Y. SINCERA: A Pipeline for Single-Cell RNA-Seq Profiling Analysis. *PLoS Comput. Biol.* **2015**, *11*, e1004575. [CrossRef]
60. Satija, R.; Farrell, J.A.; Gennert, D.; Schier, A.F.; Regev, A. Spatial reconstruction of single-cell gene expression data. *Nat. Biotechnol.* **2015**, *33*, 495–502. [CrossRef]
61. Zhao, Q.; Wang, J.; Levichkin, I.V.; Stasinopoulos, S.; Ryan, M.T.; Hoogenraad, N.J. A mitochondrial specific stress response in mammalian cells. *EMBO J.* **2002**, *21*, 4411–4419. [CrossRef]
62. Barber, R.D.; Harmer, D.W.; Coleman, R.A.; Clark, B.J. GAPDH as a housekeeping gene: Analysis of GAPDH mRNA expression in a panel of 72 human tissues. *Physiol. Genom.* **2005**, *21*, 389–395. [CrossRef]
63. Chang, Y.C.; Ding, Y.; Dong, L.; Zhu, L.J.; Jensen, R.V.; Hsiao, L.L. Differential expression patterns of housekeeping genes increase diagnostic and prognostic value in lung cancer. *PeerJ* **2018**, *6*, e4719. [CrossRef]
64. Van Der Maaten, L.; Hinton, G. Visualizing data using t-SNE. *J. Mach. Learn. Res.* **2008**, *9*, 2579–2625.

65. Herring, C.A.; Chen, B.; McKinley, E.T.; Lau, K.S. Single-Cell Computational Strategies for Lineage Reconstruction in Tissue Systems. *Cell. Mol. Gastroentero. Hepatol.* **2018**, *5*, 539–548. [CrossRef] [PubMed]
66. Amir, E.A.D.; Davis, K.L.; Tadmor, M.D.; Simonds, E.F.; Levine, J.H.; Bendall, S.C.; Shenfeld, D.K.; Krishnaswamy, S.; Nolan, G.P.; Pe'Er, D. ViSNE enables visualization of high dimensional single-cell data and reveals phenotypic heterogeneity of leukemia. *Nat. Biotechnol.* **2013**, *31*, 545–552. [CrossRef] [PubMed]
67. Becht, E.; McInnes, L.; Healy, J.; Dutertre, C.-A.; Kwok, I.W.H.; Ng, L.G.; Ginhoux, F.; Newell, E.W. Dimensionality reduction for visualizing single-cell data using UMAP. *Nat. Biotechnol.* **2018**, *37*, 38–44. [CrossRef]
68. Xiang, R.; Wang, W.; Yang, L.; Wang, S.; Xu, C.; Chen, X. A Comparison for Dimensionality Reduction Methods of Single-Cell RNA-seq Data. *Front. Genet.* **2021**, *12*, 320. [CrossRef] [PubMed]
69. Burrell, R.A.; McGranahan, N.; Bartek, J.; Swanton, C. The causes and consequences of genetic heterogeneity in cancer evolution. *Nature* **2013**, *501*, 338–345. [CrossRef] [PubMed]
70. Meacham, C.E.; Morrison, S.J. Tumour heterogeneity and cancer cell plasticity. *Nature* **2013**, *501*, 328–337. [CrossRef]
71. Navin, N.E. The first five years of single-cell cancer genomics and beyond. *Genome Res.* **2015**, *25*, 1499–1507. [CrossRef] [PubMed]
72. Hanahan, D.; Coussens, L.M. Accessories to the Crime: Functions of Cells Recruited to the Tumor Microenvironment. *Cancer Cell* **2012**, *21*, 309–322. [CrossRef] [PubMed]
73. Ben-David, U.; Siranosian, B.; Ha, G.; Tang, H.; Oren, Y.; Hinohara, K.; Strathdee, C.A.; Dempster, J.; Lyons, N.J.; Burns, R.; et al. Genetic and transcriptional evolution alters cancer cell line drug response. *Nature* **2018**, *560*, 325–330. [CrossRef] [PubMed]
74. Zhang, X.; Choi, P.S.; Francis, J.M.; Gao, G.F.; Campbell, J.D.; Ramachandran, A.; Mitsuishi, Y.; Ha, G.; Shih, J.; Vazquez, F.; et al. Somatic superenhancer duplications and hotspot mutations lead to oncogenic activation of the KLF5 transcription factor. *Cancer Discov.* **2018**, *8*, 108–125. [CrossRef] [PubMed]
75. Li, H.; Courtois, E.T.; Sengupta, D.; Tan, Y.; Chen, K.H.; Goh, J.J.L.; Kong, S.L.; Chua, C.; Hon, L.K.; Tan, W.S.; et al. Reference component analysis of single-cell transcriptomes elucidates cellular heterogeneity in human colorectal tumors. *Nat. Genet.* **2017**, *49*, 708–718. [CrossRef]
76. Park, S.R.; Namkoong, S.; Friesen, L.; Cho, C.S.; Zhang, Z.Z.; Chen, Y.C.; Yoon, E.; Kim, C.H.; Kwak, H.; Kang, H.M.; et al. Single-Cell Transcriptome Analysis of Colon Cancer Cell Response to 5-Fluorouracil-Induced DNA Damage. *Cell Rep.* **2020**, *32*, 108077. [CrossRef]
77. Valastyan, S.; Weinberg, R.A. Tumor metastasis: Molecular insights and evolving paradigms. *Cell* **2011**, *147*, 275–292. [CrossRef]
78. Klein, C.A. Parallel progression of primary tumours and metastases. *Nat. Rev. Cancer* **2009**, *9*, 302–312. [CrossRef]
79. Norton, L.; Massagué, J. Is cancer a disease of self-seeding? *Nat. Med.* **2006**, *12*, 875–878. [CrossRef]
80. Leung, M.L.; Davis, A.; Gao, R.; Casasent, A.; Wang, Y.; Sei, E.; Vilar, E.; Maru, D.; Kopetz, S.; Navin, N.E. Single-cell DNA sequencing reveals a latedissemination model in metastatic colorectal cancer. *Genome Res.* **2017**, *27*, 1287–1299. [CrossRef]
81. Tang, J.; Tu, K.; Lu, K.; Zhang, J.; Luo, K.; Jin, H.; Wang, L.; Yang, L.; Xiao, W.; Zhang, Q.; et al. Single-cell exome sequencing reveals multiple subclones in metastatic colorectal carcinoma. *Genome Med.* **2021**, *13*, 148. [CrossRef] [PubMed]
82. Wu, H.; Zhang, X.Y.; Hu, Z.; Hou, Q.; Zhang, H.; Li, Y.; Li, S.; Yue, J.; Jiang, Z.; Weissman, S.M.; et al. Evolution and heterogeneity of non-hereditary colorectal cancer revealed by single-cell exome sequencing. *Oncogene* **2017**, *36*, 2857–2867. [CrossRef] [PubMed]
83. Bian, S.; Hou, Y.; Zhou, X.; Li, X.; Yong, J.; Wang, Y.; Wang, W.; Yan, J.; Hu, B.; Guo, H.; et al. Single-cell multiomics sequencing and analyses of human colorectal cancer. *Science* **2018**, *362*, 1060–1063. [CrossRef] [PubMed]
84. Hou, Y.; Guo, H.; Cao, C.; Li, X.; Hu, B.; Zhu, P.; Wu, X.; Wen, L.; Tang, F.; Huang, Y.; et al. Single-cell triple omics sequencing reveals genetic, epigenetic, and transcriptomic heterogeneity in hepatocellular carcinomas. *Cell Res.* **2016**, *26*, 304–319. [CrossRef]
85. Pinkney, H.R.; Black, M.A.; Diermeier, S.D. Single-Cell RNA-Seq Reveals Heterogeneous lncRNA Expression in Xenografted Triple-Negative Breast Cancer Cells. *Biology* **2021**, *10*, 987. [CrossRef]
86. Schmitt, A.M.; Chang, H.Y. Long Noncoding RNAs in Cancer Pathways. *Cancer Cell* **2016**, *29*, 452–463. [CrossRef]
87. Zhang, X.; Wang, W.; Zhu, W.; Dong, J.; Cheng, Y.; Yin, Z.; Shen, F. Mechanisms and functions of long non-coding RNAs at multiple regulatory levels. *Int. J. Mol. Sci.* **2019**, *20*, 5573. [CrossRef]
88. Dykes, I.M.; Emanueli, C. Transcriptional and Post-transcriptional Gene Regulation by Long Non-coding RNA. *Genom. Proteom. Bioinform.* **2017**, *15*, 177–186. [CrossRef]
89. Liu, S.J.; Dang, H.X.; Lim, D.A.; Feng, F.Y.; Maher, C.A. Long noncoding RNAs in cancer metastasis. *Nat. Rev. Cancer* **2021**, *21*, 446–460. [CrossRef]
90. Zhang, Y.; Chen, H.; Mo, H.; Hu, X.; Gao, R.; Zhao, Y.; Liu, B.; Niu, L.; Sun, X.; Yu, X.; et al. Single-cell analyses reveal key immune cell subsets associated with response to PD-L1 blockade in triple-negative breast cancer. *Cancer Cell* **2021**. [CrossRef]
91. Hargadon, K.M.; Johnson, C.E.; Williams, C.J. Immune checkpoint blockade therapy for cancer: An overview of FDA-approved immune checkpoint inhibitors. *Int. Immunopharmacol.* **2018**, *62*, 29–39. [CrossRef]
92. Pardoll, D.M. The blockade of immune checkpoints in cancer immunotherapy. *Nat. Rev. Cancer* **2012**, *12*, 252–264. [CrossRef] [PubMed]
93. Dai, W.; Zhou, F.; Tang, D.; Lin, L.; Zou, C.; Tan, W.; Dai, Y.; Yi, K. Single-cell transcriptional profiling reveals the heterogenicity in colorectal cancer. *Medicine* **2019**, *98*, e16916. [CrossRef] [PubMed]
94. Bollen, Y.; Stelloo, E.; van Leenen, P.; van den Bos, M.; Ponsioen, B.; Lu, B.; van Roosmalen, M.J.; Bolhaqueiro, A.C.F.; Kimberley, C.; Mossner, M.; et al. Reconstructing single-cell karyotype alterations in colorectal cancer identifies punctuated and gradual diversification patterns. *Nat. Genet.* **2021**, *53*, 1187–1195. [CrossRef] [PubMed]

95. Adalsteinsson, V.A.; Tahirova, N.; Tallapragada, N.; Yao, X.; Campion, L.; Angelini, A.; Douce, T.B.; Huang, C.; Bowman, B.; Williamson, C.A.; et al. Single cells from human primary colorectal tumors exhibit polyfunctional heterogeneity in secretions of ELR+ CXC chemokines. *Integr. Biol.* **2013**, *5*, 1272–1281. [CrossRef] [PubMed]
96. Dalerba, P.; Kalisky, T.; Sahoo, D.; Rajendran, P.S.; Rothenberg, M.E.; Leyrat, A.A.; Sim, S.; Okamoto, J.; Johnston, D.M.; Qian, D.; et al. Single-cell dissection of transcriptional heterogeneity in human colon tumors. *Nat. Biotechnol.* **2011**, *29*, 1120–1127. [CrossRef]
97. Liu, M.; Liu, Y.; Di, J.; Su, Z.; Yang, H.; Jiang, B.; Wang, Z.; Zhuang, M.; Bai, F.; Su, X. Multi-region and single-cell sequencing reveal variable genomic heterogeneity in rectal cancer. *BMC Cancer* **2017**, *17*, 1–11. [CrossRef]
98. Liu, M.; Di, J.; Liu, Y.; Su, Z.; Jiang, B.; Wang, Z.; Su, X. Comparison of EpCAMhighCD44+ cancer stem cells with EpCAMhighCD44− tumor cells in colon cancer by single-cell sequencing. *Cancer Biol. Ther.* **2018**, *19*, 939–947. [CrossRef]
99. Budinska, E.; Popovici, V.; Tejpar, S.; D'Ario, G.; Lapique, N.; Sikora, K.O.; Di Narzo, A.F.; Yan, P.; Graeme Hodgson, J.; Weinrich, S.; et al. Gene expression patterns unveil a new level of molecular heterogeneity in colorectal cancer. *J. Pathol.* **2013**, *231*, 63–76. [CrossRef]
100. Riihimäki, M.; Hemminki, A.; Sundquist, J.; Hemminki, K. Patterns of metastasis in colon and rectal cancer. *Sci. Rep.* **2016**, *6*, 29765. [CrossRef]
101. Brenner, H.; Kloor, M.; Pox, C.P. Colorectal cancer. *Lancet* **2014**, *383*, 1490–1502. [CrossRef]
102. Martini, G.; Troiani, T.; Cardone, C.; Vitiello, P.; Sforza, V.; Ciardiello, D.; Napolitano, S.; Corte, C.M.D.; Morgillo, F.; Raucci, A.; et al. Present and future of metastatic colorectal cancer treatment: A review of new candidate targets. *World J. Gastroenterol.* **2017**, *23*, 4675–4688. [CrossRef] [PubMed]
103. Tamborero, D.; Rubio-Perez, C.; Deu-Pons, J.; Schroeder, M.P.; Vivancos, A.; Rovira, A.; Tusquets, I.; Albanell, J.; Rodon, J.; Tabernero, J.; et al. Cancer Genome Interpreter annotates the biological and clinical relevance of tumor alterations. *Genome Med.* **2018**, *10*, 25. [CrossRef]
104. Turner, N.C.; Reis-Filho, J.S. Genetic heterogeneity and cancer drug resistance. *Lancet Oncol.* **2012**, *13*, e178–e185. [CrossRef]
105. Dagogo-Jack, I.; Shaw, A.T. Tumour heterogeneity and resistance to cancer therapies. *Nat. Rev. Clin. Oncol.* **2018**, *15*, 81–94. [CrossRef]
106. Yates, L.R.; Campbell, P.J. Evolution of the cancer genome. *Nat. Rev. Genet.* **2012**, *13*, 795–806. [CrossRef]

medicina MDPI

Article

Benefit of Uracil–Tegafur Used as a Postoperative Adjuvant Chemotherapy for Stage IIA Colon Cancer

Po-Huang Chen [1,†], Hong-Jie Jhou [2,†], Chi-Hsiang Chung [3,4,5], Yi-Ying Wu [6], Tzu-Chuan Huang [6], Cho-Hao Lee [6], Wu-Chien Chien [3,4,7] and Jia-Hong Chen [6,*]

1 Department of Internal Medicine, Tri-Service General Hospital, National Defense Medical Center, Taipei 11490, Taiwan
2 Department of Neurology, Changhua Christian Hospital, Changhua 500, Taiwan
3 School of Public Health, National Defense Medical Center, Taipei 11490, Taiwan
4 Department of Medical Research, Tri-Service General Hospital, National Defense Medical Center, Taipei 11490, Taiwan
5 Taiwanese Injury Prevention and Safety Promotion Association (TIPSPA), Taipei 114, Taiwan
6 Division of Hematology and Oncology Medicine, Department of Internal Medicine, Tri-Service General Hospital, National Defense Medical Center, Taipei 11490, Taiwan
7 Graduate Institute of Medical Sciences, National Defense Medical Center, Taipei 11490, Taiwan
* Correspondence: ndmc_tw.tw@yahoo.com.tw (J.-H.C.); Tel.: +886-2-87923311
† These authors contributed equally to this work.

Abstract: *Background and Objectives*: Postoperative adjuvant therapy with uracil and tegafur (UFT) is often used for stage II colon cancer in Japan, but a limited number of studies have investigated the effects of UFT in these patients. *Materials and Methods*: We conducted a population-based cohort study in patients with resected stage II colon cancer comparing the outcomes after postoperative adjuvant chemotherapy with UFT with an observation-only group. The data were collected from the Taiwan National Health Insurance Research Database from 2000 to 2015. The outcomes of the study were disease-free survival (DFS) and overall survival (OS). The hazard ratios (HRs) were calculated using multivariate Cox proportional hazard regression models. Results: No differences in the DFS and OS were detected between the UFT (1137 patients) and observation (2779 patients) cohorts (DFS: adjusted HR 0.702; 95% confidence interval (CI) 0.489–1.024; $p = 0.074$) (OS: adjusted HR 0.894; 95% CI 0.542–1.186; $p = 0.477$). In the subgroup analyses of the different substages, UFT prolonged DFS in patients with stage IIA colon cancer (adjusted HR 0.652; 95% CI 0.352–0.951; $p = 0.001$) compared with DFS in the observation cohort, but no differences in the OS were detected (adjusted HR 0.734; 95% CI 0.475–1.093; $p = 0.503$). Conclusions: Our results show that DFS improved significantly in patients with stage IIA colon cancer receiving UFT as a postoperative adjuvant chemotherapy compared with DFS in the observation group.

Keywords: uracil–tegafur; colon cancer; chemotherapy; adjuvant therapy; stage IIA

check for updates

Citation: Chen, P.-H.; Jhou, H.-J.; Chung, C.-H.; Wu, Y.-Y.; Huang, T.-C.; Lee, C.-H.; Chien, W.-C.; Chen, J.-H. Benefit of Uracil–Tegafur Used as a Postoperative Adjuvant Chemotherapy for Stage IIA Colon Cancer. *Medicina* **2023**, 59, 10. https://doi.org/10.3390/medicina59010010

Academic Editors: Antonio M Scanu and Maria Rosaria De Miglio

Received: 9 November 2022
Revised: 16 December 2022
Accepted: 17 December 2022
Published: 20 December 2022

1. Introduction

Colon cancer is third in incidence and cause of cancer deaths worldwide and has been increasing rapidly in recent decades [1]. Stage II cancers have no lymph node involvement or distal metastases, and radical surgical resection of the primary tumor is the standard treatment. The prognosis after resection is relatively favorable, with a 5-year disease-free survival (DFS) rate of approximately 68%–83% after surgery alone [2]. Adjuvant therapy may be considered after surgery for patients with a high risk of recurrence to eradicate micrometastatic disease [3]. The IDEA collaboration (International Duration Evaluation of Adjuvant) recently conducted the TOSCA (Three or Six Colon Adjuvant) trial to determine the optimal duration (3 months versus 6 months) of postoperative chemotherapy in patients with high-risk stage II or III radically resected colon cancer. The conclusion showed that it was still not debatable whether 3 months of oxaliplatin-based adjuvant treatment was as efficacious as 6 months; however, the difference in survival between the two treatment durations was small [4].

For years, postoperative adjuvant chemotherapy with fluorouracil (5-FU) has been a standard of care choice among patients with locally advanced colon cancer [5]. Tegafur-uracil (UFT) is an alternative postoperative adjuvant chemotherapy. UFT is an oral drug combination of tegafur, derived from 5-FU, and uracil in a molar ratio of 1:4. UFT acts as a competitive inhibitor of dihydropyrimidine dehydrogenase (DPD) [6]. Because of UFT's tolerability and safety in an outpatient setting, it is the most commonly prescribed adjuvant chemotherapeutic for colon cancer in Taiwan [7]. A nationwide cohort study and meta-analysis demonstrated the similar effects of UFT and intravenous 5-FU on DFS and overall survival (OS), but UFT has a lower incidence of adverse events when used as a postoperative adjuvant chemotherapy in patients with locally advanced colon cancer [8].

Adjuvant chemotherapy represents a dilemma for clinical oncologists when treating patients with stage II colon cancer receiving surgical resection. In previous studies, the survival benefits of adjuvant chemotherapy were still inconclusive in unselected stage II patients. The adjuvant therapy benefits may be limited in patients with average-risk stage II cancer and a relatively good prognosis. According to the 2021 American Society of Clinical Oncology guidelines, adjuvant chemotherapy is not recommended for patients with a low risk of recurrence [9]. Thus, high-risk features should be identified to determine which subgroups might benefit from adjuvant chemotherapy. The current guidelines suggest that patients with one or more high-risk features should receive adjuvant chemotherapy, including pT4, bowel obstruction, or tumor perforation; fewer than 12 lymph nodes harvested; vascular, lymphatic, or perineural invasion; and a poorly differentiated histology [2]. However, the relative prognostic weight of these features is not considered.

Although postoperative adjuvant therapy with UFT is often used for stage II colon cancer in Japan, a limited number of studies have investigated the effects of UFT in these patients. Thus, clarifying whether postoperative adjuvant treatment with UFT is beneficial in stage II colon cancer patients at different substages compared with observation alone in a larger population, and a real-world setting is needed. This study, using a population-based database, aimed to examine the survival benefit of oral UFT compared with observation only for postoperative stage II colon cancer patients analyzed at different substages.

2. Materials and Methods

2.1. Data Source

The data were gathered from databases provided by the Health and Welfare Data Science Center, including the Taiwan Cancer Registry (TCR) Database 2000–2015 [10] and the National Health Insurance Research Database (NHIRD). 2000–2015.

The TCR is organized and funded by the Ministry of Health and Welfare and is managed by the Taiwan Public Health Association. All hospitals in Taiwan with at least 50 beds are required to report all newly diagnosed and confirmed malignancies to the registry. Detailed information on diagnosis, treatments, and outcomes is collected from 80 hospitals, covering more than 90% of all cancer cases diagnosed annually in Taiwan. Diagnoses are coded according to the International Classification of Diseases for Oncology, 3rd Edition, format [11].

A nationwide population-based study was conducted using data from 2000 to 2015 obtained from the Longitudinal Health Insurance Database (LHID) of Taiwan [12]. Two million beneficiaries from the NHIRD registry were randomly sampled. The LHID includes the following claims data: sociodemographic information, medical visits, emergency care, hospitalization, surgical procedure, medication, and other medical services. Diseases are diagnosed according to the International Classification of Diseases, Ninth Revision, Clinical Modification (ICD-9-CM) codes. The Registry for Catastrophic Illness Patient Database (RCIPD) includes data from insured residents with severe diseases, such as malignancies, as defined by the NHI program [13].

This study was approved by the Institutional Review Board of the Tri-Service General Hospital, Taipei, Taiwan (TSGHIRB No. B-110-12). Because the data from the NHI were de-identified, the signed informed consent of the included patients was waived.

2.2. Study Population and Definition of Statin Exposure

Patients newly diagnosed with colon cancer (ICD-9 Code: 153–154.1) from 1 January 2000 to 31 December 2015, were identified from the NHIRD database. The malignancy diagnosis was confirmed using the RCIPD data. Patients with stage II colon cancer who received operative therapy within 6 months after the diagnosis (ICD-9-CM Procedure Code: OP 45.21, OP 45.71–45.76, OP 45.79, OP 45.8, and OP 48.4–48.6) were identified from the TCR database, and their data were retrieved. The TCR uses the American Joint Committee on Cancer staging system, 7th Edition, to record the stages of all cancer patients. We excluded patients that received other target or chemotherapies, including bevacizumab, cetuximab, capecitabine, irinotecan, oxaliplatin, or 5-FU; diagnosed cancer before the index date using ICD-9-CM140–239; diagnosed secondary malignancy (ICD-9-CM: 196–198.9); or benign colon neoplasm (ICD-9-CM: 211.3 and 211.4). The UFT cohort comprised patients who were prescribed UFT, and the observation (without treatment) cohort comprised patients who did not receive any postoperative chemotherapy.

2.3. Outcome and Comorbidities Measurement

The outcomes of interest were the DFS and the OS. DFS was defined as the time interval from the first day postoperation to tumor recurrence or death. OS was defined as the time interval from the first day postoperation to death. We also extracted the covariates of the patients, including age, sex, stage, and underlying diseases. The Charlson Comorbidity Index (CCI) categorizes the comorbidities of patients based on the ICD codes [14]. The CCI Revised (CCI_R) was calculated by removing the variables mentioned and accounted for in the baseline comorbidities. The socioeconomic status of the study participants was approximated using insurance premiums (i.e., income level), level of care (stratified by the levels of hospital, including central, regional, or local hospitals determined by the Taiwanese government), and urbanization levels [15]. Additional analyses were conducted to ascertain the impact of adjuvant chemotherapy on DFS and OS in patients at different substages (stage IIA/IIB/IIC) in the UFT and observation groups.

2.4. Statistical Analysis

The categorical variables are expressed using numbers (i.e., percentages), and the continuous variables are expressed as the mean ± standard deviations (SDs). The chi-square test or Fisher's exact test was used to compare the categorical variables, whereas t-tests were used to compare the mean difference for continuous variables among the UFT and observation groups. Univariate and multivariate Cox regression analyses were employed to evaluate the crude and adjusted hazard ratios (HRs) for the influence (odds) of the analyzed variables on DFS and OS; the observation group was used as a reference. We adjusted the multivariate Cox regression model using all of the characteristics, including age; sex; insurance premium; level of care; urbanization; comorbidities, including hypertension, diabetes mellitus (DM), chronic obstructive pulmonary disease (COPD), chronic kidney disease (CKD), ischemic heart disease (IHD), congestive heart disease (CHD), and stroke; and a CCI_R. Kaplan–Meier analysis and log-rank tests of DFS and OS based on the stage of colon cancer were performed. The two-sided p-values of the log-rank test less than 0.05 were considered statistically significant. All statistical analyses were performed using IBM SPSS Statistics for Windows version 22.0 (IBM Corp., Armonk, NY, USA).

3. Results

A total of 3916 surgical patients with stage II colon cancer were observed in this study, including 1137 patients in the UFT group and 2779 patients in the observation group. Figure 1 shows a flow chart of the recruitment of subjects from the NHIRD.

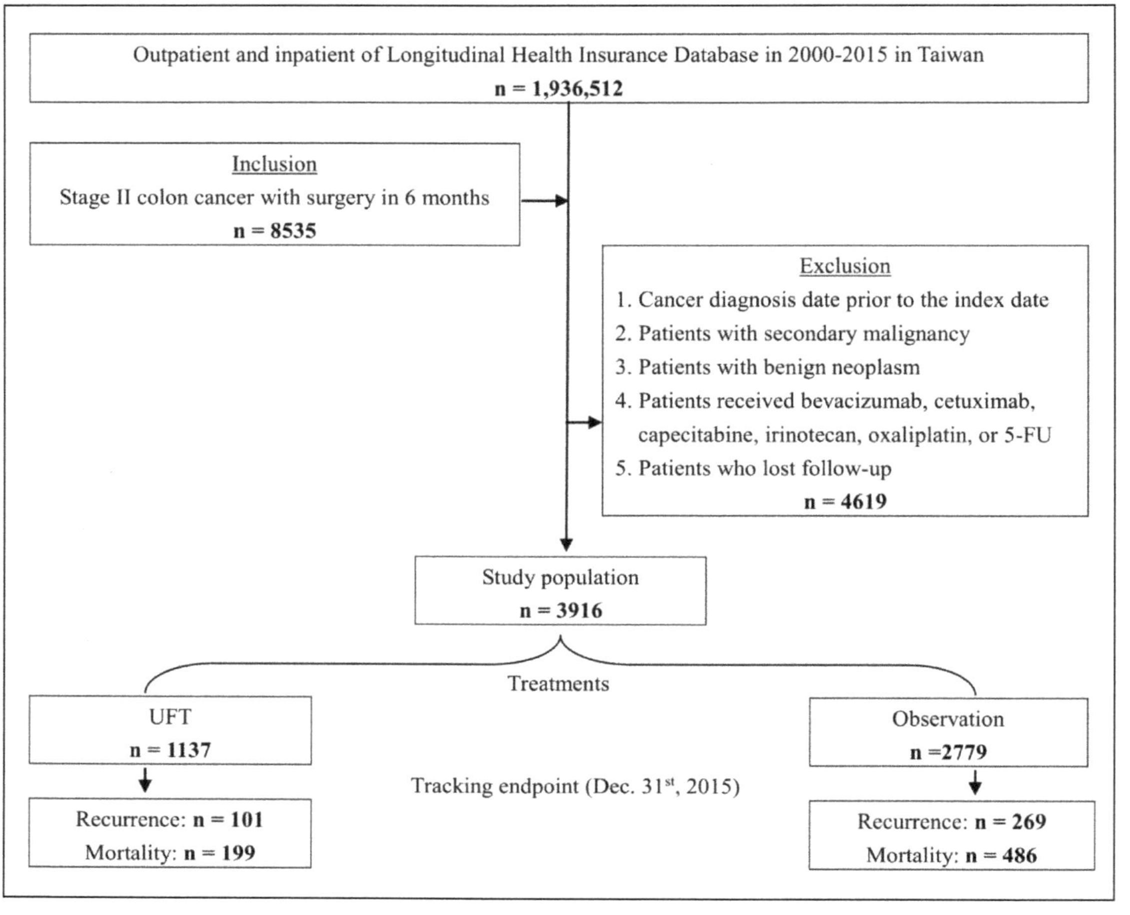

Figure 1. A flow chart of the recruitment of the subjects.

3.1. Patient Characteristics

Table 1 shows the characteristics and baseline comorbidity status of the UFT (n = 1137) and observation (n = 2779) cohorts. The percentages of males in the UFT and observation cohorts were 59.89% and 58.94%, respectively. The mean ± SD ages for the UFT and observation cohorts were 63.40 ± 10.25 and 65.12 ± 11.12 years, respectively. The mean follow-up periods were 7.20 ± 6.84 and 7.22 ± 6.87 years in the UFT and observation cohorts, respectively. No significant differences in sex; age; insured premium; urbanization; or comorbidities, including hypertension, DM, COPD, CKD, IHD, CHD, and stroke; or CCI_R index were detected (the social–economic data are shown in Supplementary Materials Table S1; the follow-up period data are shown in Supplementary Materials Table S2A,B; the ICD-9-CM, NHI code, and definition are shown in Supplementary Material Table S3).

Table 1. Characteristics of study in the baseline.

Treatment Variables		UFT		Observation		*p*-Value
		n	%	n	%	
Total		1137	29.03	2779	70.97	
Gender						0.582
Male		681	59.89	1638	58.94	
Female		456	40.11	1141	41.06	
Age (years ± SD)		63.40 ± 10.25		65.12 ± 11.12		<0.001
HTN	With	330	29.02	781	28.10	0.562
	Without	807	70.98	1998	71.90	
DM	With	201	17.68	455	16.37	0.321
	Without	936	82.32	2324	83.63	
COPD	With	55	4.84	108	3.89	0.186
	Without	1082	95.16	2671	96.11	
CKD	With	17	1.50	36	1.30	0.648
	Without	1120	98.50	2743	98.70	
IHD	With	60	5.28	145	5.22	0.937
	Without	1077	94.72	2634	94.78	
CHD	With	24	2.11	54	1.94	0.733
	Without	1113	97.89	2725	98.06	
Stroke	With	33	2.90	80	2.88	0.968
	Without	1104	97.10	2699	97.12	
CCI_R		1.03 ± 0.19		1.03 ± 0.15		0.998

p-Value: categorical variables: chi-squared/Fisher's exact test; continuous variables: *t*-test. UFT, uracil–tegafur; HTN, hypertension; DM, diabetes mellitus; COPD, chronic obstructive pulmonary disease; CKD, chronic kidney disease; IHD, ischemic heart disease; CHD, congestive heart disease; CCI_R, Charlson comorbidity index revised.

3.2. Disease-Free Survival

The Kaplan–Meier plots with log-rank tests revealed significant differences in DFS between the UFT and observation cohorts (log-rank test: $p < 0.001$; Figure 2). According to the multivariate Cox regression model, DFS did not differ significantly between the UFT and observation groups (UFT vs. observation; adjusted HR 0.702; 95% CI 0.489–1.024; $p = 0.074$; Table 2). Male sex, having comorbidities (i.e., HTN, DM, CKD, IHD, CHD, and stroke), and the influence of the CCI_R score were significant factors with shorter DFS.

3.3. Overall Survival

The Kaplan–Meier plots with log-rank tests revealed significant differences in the OS between the UFT and observation cohorts (log-rank test: $p < 0.001$; Figure 3).

The multivariate Cox regression model indicated that the OS did not differ significantly between the UFT and observation groups (adjusted HR 0.894; 95% CI 0.542–1.186; $p = 0.477$; Table 2). Male sex, older age, having comorbidities (i.e., HTN, DM, CKD, IHD, CHD, and stroke), and the influence of the CCI_R score were significant factors with shorter OS.

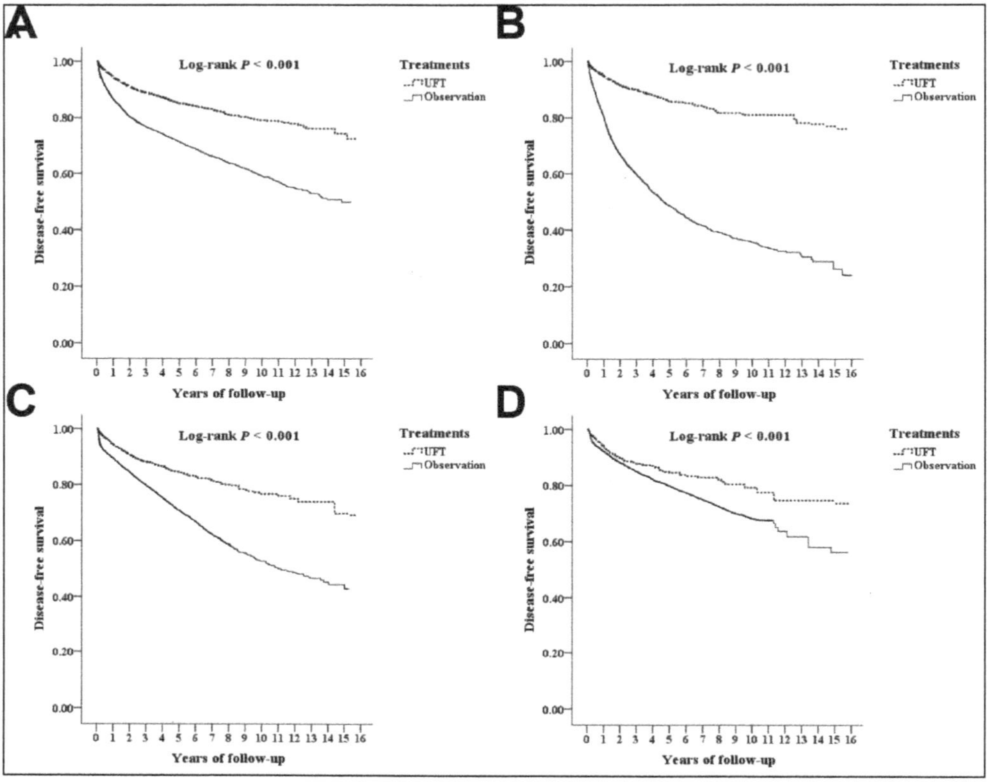

Figure 2. Kaplan–Meier plots for cumulative risk in disease-free survival: overall populations (**A**); Stage IIA (**B**); Stage IIB (**C**); Stage IIC (**D**).

3.4. Analyses for the Different Substages

Among the patients with stage IIA colon cancer, the Kaplan–Meier plots with log-rank tests revealed significant differences in the DFS (log-rank test: $p < 0.001$; Figure 2) and OS (log-rank test: $p < 0.001$; Figure 3) between the UFT and observation cohorts. The multivariate Cox regression model indicated that the DFS increased significantly in patients with stage IIA colon cancer receiving postoperative UFT adjuvant chemotherapy compared with the DFS in the observation group (adjusted HR 0.652; 95% CI 0.352–0.951; $p = 0.001$; Table 3); however, no differences in the OS were detected (adjusted HR 0.734; 95% CI 0.475–1.093; $p = 0.503$; Table 3)

In patients with stages IIB and IIC colon cancer, the Kaplan–Meier plots with log-rank tests revealed significant differences in the DFS (both $p < 0.001$; Figure 2) and OS (both $p < 0.001$; Figure 3) between the UFT and observation cohorts. According to the multivariate Cox regression model in the patients with stages IIB and IIC colon cancer, no difference in the DFS and OS between the UFT and observation groups were detected (DFS: stage IIB, adjusted HR 0.713, 95% CI 0.492–1.029, $p = 0.079$; DFS: stage IIC, adjusted HR 0.804, 95% CI 0.575–1.125, $p = 0.184$; Table 2) (OS: stage IIB, adjusted HR 0.877; 95% CI 0.531–1.153, $p = 0.483$; OS: stage IIC, adjusted HR 0.904, 95% CI 0.638–1.256, $p = 0.425$; Table 3).

Table 2. Cox regression analysis of disease-free survival and overall survival.

Prognosis	Disease-Free Survival (DFS)				Overall Survival (OS)			
Variables	**Crude HR (95% CI)**	*p*	**Adjusted HR (95% CI)**	*p*	**Crude HR (95% CI)**	*p*	**Adjusted HR (95% CI)**	*p*
Treatments								
UFT vs. observation	0.613 (0.377–0.806)	<0.001	0.702 (0.489–1.024)	0.074	0.785 (0.426–0.894)	<0.001	0.894 (0.542–1.186)	0.477
Gender								
Male vs. female	1.365 (1.124–1.503)	<0.001	1.265 (1.106 –1.482)	<0.001	1.489 (1.303 –1.677)	<0.001	1.420 (1.298–1.583)	<0.001
Age Groups (Years)								
<30	Reference							
30–39	1.124 (0.822–1.825)	0.182	1.024 (0.724–1.781)	0.389	2.561 (1.786 –4.486)	<0.001	2.008 (1.025–3.349)	0.035
40–49	1.304 (0.913–1.911)	0.094	1.203 (0.902–1.924)	0.172	3.789 (2.229–5.702)	<0.001	2.186 (1.097–3.570)	0.001
50–59	1.386 (0.972–1.934)	0.067	1.186 (0.851–1.876)	0.234	5.978 (3.224–9.972)	<0.001	4.299 (2.004–8.301)	<0.001
≧60	1.402 (1.020–2.020)	0.030	1.354 (0.989–1.986)	0.069	7.124 (4.809–13.312)	<0.001	5.038 (2.897–9.896)	<0.001
HTN	1.678 (1.307–1.882)	<0.001	1.562 (1.265–1.782)	<0.001	1.863 (1.511–2.104)	<0.001	1.782 (1.428–2.006)	<0.001
DM	1.831 (1.367–2.010)	<0.001	1.762 (1.303–1.977)	<0.001	2.030 (1.724–2.308)	<0.001	1.975 (1.629–2.210)	<0.001
COPD	1.382 (0.986–1.769)	0.072	1.283 (0.865–1.677)	0.277	1.397 (1.002–1.784)	0.049	1.270 (0.852–1.624)	0.289
CKD	1.482 (1.153–1.780)	<0.001	1.293 (1.021–1.445)	0.029	2.156 (1.503–2.970)	<0.001	2.011 (1.452–2.897)	<0.001
IHD	1.686 (1.112–1.897)	<0.001	1.553 (1.086–1.795)	0.002	1.918 (1.628–2.774)	<0.001	1.897 (1.583–2.610)	<0.001
CHD	1.735 (1.442–1.975)	<0.001	1.652 (1.352–1.896)	<0.001	1.993 (1.586–2.601)	<0.001	1.824 (1.550–2.533)	<0.001
Stroke	1.808 (1.553–2.030)	<0.001	1.771 (1.448–1.909)	<0.001	2.030 (1.724–2.789)	<0.001	1.902 (1.652–2.672)	<0.001
CCI_R	1.304 (1.205–1.488)	<0.001	1.246 (1.112–1.304)	<0.001	1.372 (1.289–1.483)	<0.001	1.297 (1.158–1.372)	<0.001

HR, hazard ratio; CI, confidence interval; adjusted HR, adjusted variables listed in the table.

Table 3. Factors of prognosis stratified by cancer stage.

UFT vs. Observation		Disease-Free Survival (DFS)			
Stage	**Patients**	**Adjusted HR**	**95% CI**	**95% CI**	*p*
Overall	3916	0.702	0.489	1.024	0.074
Stage IIA	2326	0.652	0.352	0.951	0.001
Stage IIB	819	0.713	0.492	1.029	0.079
Stage IIC	871	0.804	0.575	1.125	0.184
		Overall Survival (OS)			
Stage	**Patients**	**Adjusted HR**	**95% CI**	**95% CI**	*p*
Overall	3916	0.894	0.542	1.186	0.477
Stage IIA	2326	0.734	0.475	1.093	0.503
Stage IIB	819	0.877	0.531	1.153	0.483
Stage IIC	871	0.904	0.638	1.256	0.425

Adjusted HR, adjusted hazard ratio (adjusted for the variables listed in Table 2); CI, confidence interval.

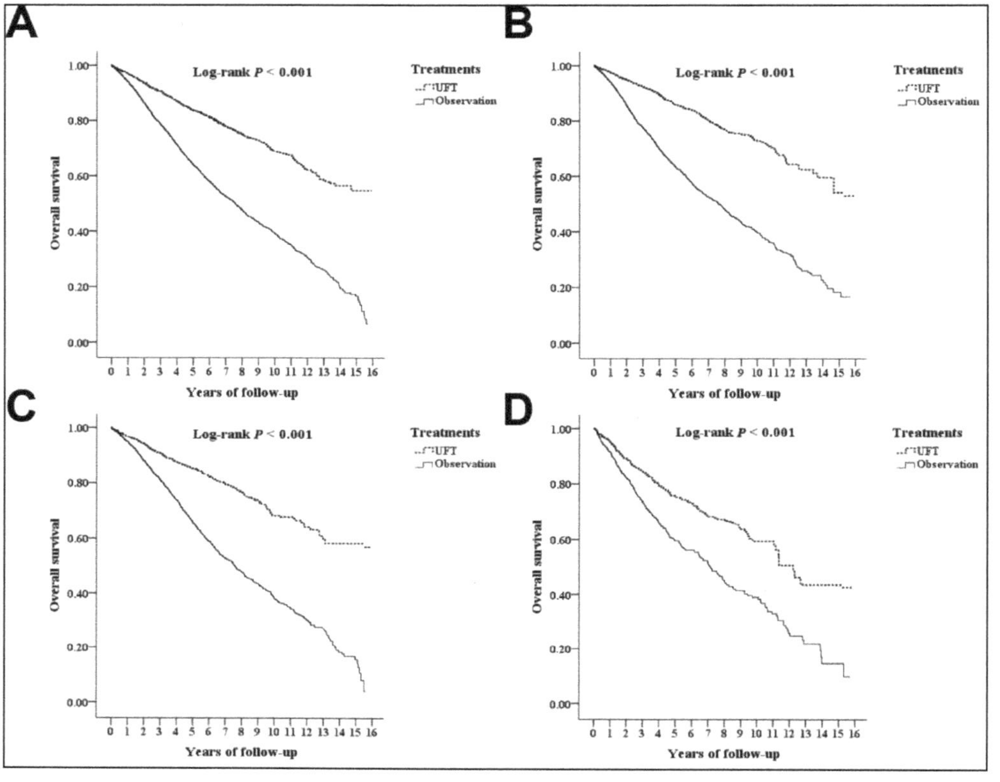

Figure 3. Kaplan–Meier plots for the cumulative risk in overall survival: overall populations (**A**); Stage IIA (**B**); Stage IIB (**C**); Stage IIC (**D**).

4. Discussion

This nationwide, large-scale, retrospective cohort study compared the effectiveness of postoperative adjuvant chemotherapy with UFT to observation only in stage II colon cancer patients. In the 15-year follow-up cohorts, UFT showed no difference with observation only in DFS and OS. However, in the subgroup analysis of stage IIA, the patients who received UFT as a postoperative adjuvant chemotherapy had significantly prolonged DFS compared with observation alone.

4.1. UFT Effectiveness

Currently, 5-FU (5-FU/LV, capecitabine, UFT, and S-1) and oxaliplatin are the main drugs used as postoperative adjuvant chemotherapy for colon cancer. The usual dosage of intravenous 5-FU is a weekly 24 hour infusion of a maximal tolerable dose of 5-FU (2600 mg/m^2) and LV (500 mg/m^2) for 6 months [16]. Two oral UFT capsules, containing tegafur 100 mg and uracil 224 mg, are administered twice per day (400 mg of tegafur per day) in Taiwan [17]. A 5-day treatment plus a 2-day rest regimen of UFT for 12 months was beneficial in the NSAS-CC study [18]. In contrast, the NSABP C-06 study showed that oral UFT had DFS and OS similar to those of intravenous 5-FU/LV. However, patients treated with UFT had a better quality of life than those treated with 5-FU/LV [19].

4.2. Survival Paradox in Patients with Localized Advanced Colon Cancer

A survival paradox was noted between stage IIB/C (T4N0) and stage IIIA (T1-2N1 and T1N2a) colon cancer in previous studies [20–23]. Li et al. [24] found that the colon-cancer-specific survival (CCSS) rate of the stage IIIA colon cancer patients were significantly higher than that of the stage IIB and IIC colon cancer patients (5-year CCSS rates for stage IIB vs. stage IIC vs. stage IIIA: 74.2% vs. 72.5% vs. 91.9%). Furthermore, Mo et al. analyzed data from the US Surveillance, Epidemiology, and End Results (SEER) database, and showed that patients with stage IIA rectal cancer had worse survival than patients with stage IIIA disease [25]. The inferior survival in stage II compared with stage IIIA may be due to the lower use of systemic chemotherapy in stage II colon cancer patients. Thus, we should improve the survival of stage II colon cancer patients, especially in stage IIA. The 5-year DFS improved in patients with low-risk IIA colon cancer after receiving adjuvant chemotherapy with UFT more than 12 months after surgery in a retrospective cohort study [26], which is consistent with our finding.

4.3. Survival Risk Factors in Patients with Localized Advanced Colon Cancer

Over the past decades, subgroup analyses from large adjuvant trials have investigated the impacts of risk factors in prognosis [27]. The crucial prognostic factors for disease progression included T stage as tumor size, lymph node status, pathological grading, and microsatellite status.

The T stage is highly associated with tumor size. In a study conducted by Mo et al. [25], the mean tumor size of stage IIA rectal cancer was larger than the tumor size of stage IIIA rectal cancer in both the SEER and FUSCC cohorts. Therefore, en bloc resection of T1/T2 tumors may be much easier to achieve than the resection of T3/T4 tumors, making surgical negative margins more difficult to achieve for a high T level stage IIA colon cancer than a low T level IIIA disease. The larger tumor size in stage IIA colon cancer can increase the surgical margin positivity, thus escalating the recurrence rate of colon cancer and jeopardizing the prognosis of colon cancer patients.

Lymph node status is a crucial prognostic factor in colorectal cancer to determine postoperative managements and follow-up plans [28,29]. Stage III is distinguished from stage II colorectal cancer by the presence of lymph node metastases. According to the recommendation by the American Joint Committee on Cancer and the College of American Pathologists, at least 12 lymph nodes should be examined to adequately stage colorectal cancer patients [30,31]. Examining an adequate number of lymph nodes has been regarded as a key quality measure for colon cancer care in the United States since 2006 [32].

A high grade, indicating poorly differentiated disease, is associated with poor prognosis [33]. In a cohort of 3302 stage II and stage III colon cancer patients, Gill et al. observed lower 5-year DFS and OS in high-grade disease. In addition, high-grade disease was related to a loss of 8%–9% in 5-year DFS in T3N0 and T4N0 tumors compared with low-grade disease (65% vs. 73% and 51% vs. 60%, respectively) [34].

Two groups of colorectal cancers can be distinguished based on the state of mismatch repair: MSI-high (MSI-H, deficiency of the mismatch repair) and MSI-low (proficiency of the mismatch repair, pMMR). Adjuvant chemotherapy is not suggested for MSI-H stage II patients without high-risk features; therefore, observation is considered a reasonable treatment option [35].

4.4. Limitations

Several limitations should be considered when interpreting the results of this study. First, the NHIRD had insufficiently detailed clinical data, including the severity of lymph node involvement, pathologic grade, microsatellite instability status, the reasons for each patient's treatment plan, or the quality of surgery. Second, patients with cancer were defined using claims data and diagnostic codes. The diagnostic accuracy remained unclear, and disease misclassification might cause false associations [36]. Third, the National Quality Forum has listed the assessment of at least 12 lymph nodes among the key quality measures for colon cancer care in the United States since 2006. However, data for this study were from 2000 to 2015; hence, the number of lymph nodes examined may be insufficient. Finally, potential confounders may have occurred that could bias the results. There remains a need to perform further analysis using the clinical data from individual participants and controlling for potential confounders [37].

To the best of our knowledge, our study is the first real-world study to examine the effectiveness of UFT in stage II colon cancer and its substages and a large-scale study to strengthen the statistical power [38]. A stratified analysis was conducted using demographic characteristics, including age, sex, socioeconomic status, and comorbidities, to exam the clinical heterogenicity.

5. Conclusions

Our results show that the DFS improved significantly in patients with stage IIA colon cancer receiving UFT as a postoperative adjuvant chemotherapy compared with the DFS in the observation group.

Supplementary Materials: The following are available online at https://www.mdpi.com/article/10.3390/medicina59010010/s1.

Author Contributions: Conceptualization, P.-H.C., H.-J.J., C.-H.C. and W.-C.C.; data curation, P.-H.C., C.-H.L. and W.-C.C.; formal analysis, C.-H.C., Y.-Y.W. and T.-C.H.; funding acquisition, Jia-Hong Chen; investigation, P.-H.C., C.-H.C. and J.-H.C.; methodology, H.-J.J. and C.-H.C.; project administration, W.-C.C. and J.-H.C.; resources, Y.-Y.W., T.-C.H. and J.-H.C.; software, C.-H.C.; supervision, H.-J.J. and C.-H.L.; validation, T.-C.H.; writing—original draft, P.-H.C. and H.-J.J.; writing—review and editing, Y.-Y.W., C.-H.L., W.-C.C. and J.-H.C. All authors have read and agreed to the published version of the manuscript.

Funding: This study was supported by the Tri-Service General Hospital Research Foundation (No. TSGH-D-112157), and the sponsor has no role in study design, data collection and analysis, decision to publish, or preparation of the manuscript.

Institutional Review Board Statement: The study was conducted in accordance with the Declaration of Helsinki, and approved by the Institutional Review Board of Tri-Service General Hospital (TSGHIRB No. B-109–11, Approve date: 2020/05/11).

Informed Consent Statement: Patient consent was waived due to all NHIRD data being anonymous and de-identified.

Data Availability Statement: Data are available from the National Health Insurance Research Database (NHIRD) published by Taiwan National Health Insurance (NHI) Bureau. Due to legal restrictions imposed by the government of Taiwan in relation to the "Personal Information Protection Act", data cannot be made publicly available. Requests for data can be sent as a formal proposal to the NHIRD (http://nhird.nhri.org.tw).

Ethics approval: The investigational protocols were approved by the official peer review committee in the Tri-Service General Hospital, and the protocol number was TSGHIRB No. B-109–11 (Approval date: 11 May 2020).

Acknowledgments: The authors would like to thank Enago (www.enago.tw, accessed on 8 November 2022) and MDPI English Editing for the English language review.

Conflicts of Interest: The authors declare no conflict of interest.

References

1. Siegel, R.L.; Miller, K.D.; Goding Sauer, A.; Fedewa, S.A.; Butterly, L.F.; Anderson, J.C.; Cercek, A.; Smith, R.A.; Jemal, A. Colorectal cancer statistics, 2020. *CA A Cancer J. Clin.* **2020**, *70*, 145–164. [CrossRef] [PubMed]
2. Argilés, G.; Tabernero, J.; Labianca, R.; Hochhauser, D.; Salazar, R.; Iveson, T.; Laurent-Puig, P.; Quirke, P.; Yoshino, T.; Taieb, J.; et al. Localised colon cancer: ESMO Clinical Practice Guidelines for diagnosis, treatment and follow-up. *Ann. Oncol.* **2020**, *31*, 1291–1305. [CrossRef] [PubMed]
3. Chang, C.L.; Yuan, K.S.; Wu, A.T.H.; Wu, S.Y. Adjuvant Therapy for High-Risk Stage II or III Colon Adenocarcinoma: A Propensity Score-Matched, Nationwide, Population-Based Cohort Study. *Cancers* **2019**, *11*, 2003. [CrossRef] [PubMed]

4. Lonardi, S.; Sobrero, A.; Rosati, G.; Di Bartolomeo, M.; Ronzoni, M.; Aprile, G.; Scartozzi, M.; Banzi, M.; Zampino, M.G.; Pasini, F.; et al. Phase III trial comparing 3–6 months of adjuvant FOLFOX4/XELOX in stage II–III colon cancer: Safety and compliance in the TOSCA trial. *Ann. Oncol.* **2016**, *27*, 2074–2081. [CrossRef] [PubMed]
5. Wolmark, N.; Rockette, H.; Mamounas, E.; Jones, J.; Wieand, S.; Wickerham, D.L.; Bear, H.D.; Atkins, J.N.; Dimitrov, N.V.; Glass, A.G.; et al. Clinical trial to assess the relative efficacy of fluorouracil and leucovorin, fluorouracil and levamisole, and fluorouracil, leucovorin, and levamisole in patients with Dukes' B and C carcinoma of the colon: Results from National Surgical Adjuvant Breast and Bowel Project C-04. *J. Clin. Oncol.* **1999**, *17*, 3553–3559. [CrossRef]
6. Lin, J.K.; Wang, W.S.; Hsieh, R.K.; Hsu, T.C.; Chiou, T.J.; Liu, J.H.; Fan, F.S.; Yen, C.C.; Lin, T.C.; Jiang, J.K.; et al. Phase II study of oral tegafur-uracil and folinic acid as first-line therapy for metastatic colorectal cancer: Taiwan experience. *Jpn. J. Clin. Oncol.* **2000**, *30*, 510–514. [CrossRef]
7. Huang, W.Y.; Ho, C.L.; Lee, C.C.; Hsiao, C.W.; Wu, C.C.; Jao, S.W.; Yang, J.F.; Lo, C.H.; Chen, J.H. Oral tegafur-uracil as metronomic therapy following intravenous FOLFOX for stage III colon cancer. *PLoS ONE* **2017**, *12*, e0174280. [CrossRef]
8. Chen, P.H.; Wu, Y.Y.; Lee, C.H.; Chung, C.H.; Chen, Y.G.; Huang, T.C.; Yeh, R.H.; Chang, P.Y.; Dai, M.S.; Lai, S.W.; et al. Uracil-tegafur vs fluorouracil as postoperative adjuvant chemotherapy in Stage II and III colon cancer: A nationwide cohort study and meta-analysis. *Medicine* **2021**, *100*, e25756. [CrossRef]
9. Baxter, N.N.; Kennedy, E.B.; Bergsland, E.; Berlin, J.; George, T.J.; Gill, S.; Gold, P.J.; Hantel, A.; Jones, L.; Lieu, C.; et al. Adjuvant Therapy for Stage II Colon Cancer: ASCO Guideline Update. *J. Clin. Oncol.* **2022**, *40*, 892–910. [CrossRef]
10. Kao, C.W.; Chiang, C.J.; Lin, L.J.; Huang, C.W.; Lee, W.C.; Lee, M.Y. Accuracy of long-form data in the Taiwan cancer registry. *J. Formos. Med. Assoc.* **2021**, *120*, 2037–2041. [CrossRef]
11. Chiang, C.J.; You, S.L.; Chen, C.J.; Yang, Y.W.; Lo, W.C.; Lai, M.S. Quality assessment and improvement of nationwide cancer registration system in Taiwan: A review. *Jpn. J. Clin. Oncol.* **2015**, *45*, 291–296. [CrossRef]
12. Hsieh, C.Y.; Su, C.C.; Shao, S.C.; Sung, S.F.; Lin, S.J.; Kao Yang, Y.H.; Lai, E.C. Taiwan's National Health Insurance Research Database: Past and future. *Clin. Epidemiol.* **2019**, *11*, 349–358. [CrossRef] [PubMed]
13. Sung, S.-F.; Hsieh, C.-Y.; Hu, Y.-H. Two Decades of Research Using Taiwan's National Health Insurance Claims Data: Bibliometric and Text Mining Analysis on PubMed. *J. Med. Internet. Res.* **2020**, *22*, e18457. [CrossRef]
14. Simard, M.; Sirois, C.; Candas, B. Validation of the Combined Comorbidity Index of Charlson and Elixhauser to Predict 30-Day Mortality Across ICD-9 and ICD-10. *Med. Care* **2018**, *56*, 441–447. [CrossRef] [PubMed]
15. Huang, N.C.; Kung, S.F.; Hu, S.C. The Relationship between Urbanization, the Built Environment, and Physical Activity among Older Adults in Taiwan. *Int. J. Environ. Res. Public Health* **2018**, *15*, 836. [CrossRef] [PubMed]
16. Chen, L.T.; Whang-Peng, J. Current status of clinical studies for colorectal cancer in Taiwan. *Clin. Color. Cancer* **2004**, *4*, 196–203. [CrossRef] [PubMed]
17. Chen, T.C.; Jeng, Y.M.; Liang, J.T. Metronomic chemotherapy with tegafur-uracil following radical resection in stage II colorectal cancer. *J. Formos. Med. Assoc.* **2021**, *120*, 1194–1201. [CrossRef] [PubMed]
18. Hamaguchi, T.; Shirao, K.; Moriya, Y.; Yoshida, S.; Kodaira, S.; Ohashi, Y. Final results of randomized trials by the National Surgical Adjuvant Study of Colorectal Cancer (NSAS-CC). *Cancer Chemother. Pharmacol.* **2011**, *67*, 587–596. [CrossRef] [PubMed]
19. Lembersky, B.C.; Wieand, H.S.; Petrelli, N.J.; O'Connell, M.J.; Colangelo, L.H.; Smith, R.E.; Seay, T.E.; Giguere, J.K.; Marshall, M.E.; Jacobs, A.D.; et al. Oral uracil and tegafur plus leucovorin compared with intravenous fluorouracil and leucovorin in stage II and III carcinoma of the colon: Results from National Surgical Adjuvant Breast and Bowel Project Protocol C-06. *J. Clin. Oncol.* **2006**, *24*, 2059–2064. [CrossRef] [PubMed]
20. Chu, Q.D.; Zhou, M.; Medeiros, K.L.; Peddi, P.; Kavanaugh, M.; Wu, X.C. Poor survival in stage IIB/C (T4N0) compared to stage IIIA (T1-2 N1, T1N2a) colon cancer persists even after adjusting for adequate lymph nodes retrieved and receipt of adjuvant chemotherapy. *BMC Cancer* **2016**, *16*, 460. [CrossRef] [PubMed]
21. Kim, M.J.; Jeong, S.Y.; Choi, S.J.; Ryoo, S.B.; Park, J.W.; Park, K.J.; Oh, J.H.; Kang, S.B.; Park, H.C.; Heo, S.C.; et al. Survival paradox between stage IIB/C (T4N0) and stage IIIA (T1-2N1) colon cancer. *Ann. Surg. Oncol.* **2015**, *22*, 505–512. [CrossRef] [PubMed]
22. Chu, Q.D.; Zhou, M.; Medeiros, K.; Peddi, P. Positive surgical margins contribute to the survival paradox between patients with stage IIB/C (T4N0) and stage IIIA (T1-2N1, T1N2a) colon cancer. *Surgery* **2016**, *160*, 1333–1343. [CrossRef] [PubMed]
23. Kim, H.S.; Kim, K.M.; Lee, S.B.; Kim, G.R.; Han, Y.D.; Cho, M.S.; Hur, H.; Lee, K.Y.; Kim, N.K.; Min, B.S. Clinicopathological and biomolecular characteristics of stage IIB/IIC and stage IIIA colon cancer: Insight into the survival paradox. *J. Surg. Oncol.* **2019**, *120*, 423–430. [CrossRef] [PubMed]
24. Li, H.; Fu, G.; Wei, W.; Huang, Y.; Wang, Z.; Liang, T.; Tian, S.; Chen, H.; Zhang, W. Re-Evaluation of the Survival Paradox Between Stage IIB/IIC and Stage IIIA Colon Cancer. *Front. Oncol.* **2020**, *10*, 595107. [CrossRef]
25. Mo, S.; Dai, W.; Xiang, W.; Huang, B.; Li, Y.; Feng, Y.; Li, Q.; Cai, G. Survival Contradiction Between Stage IIA and Stage IIIA Rectal Cancer: A Retrospective Study. *J. Cancer* **2018**, *9*, 1466–1475. [CrossRef]
26. Hu, J.M.; Chou, Y.C.; Wu, C.C.; Hsiao, C.W.; Lee, C.C.; Chen, C.T.; Hu, S.I.; Liu, W.T.; Jao, S.W. Adjuvant chemotherapy with tegafur/uracil for more than 1 year improves disease-free survival for low-risk Stage II colon cancer. *J. Chin. Med. Assoc.* **2016**, *79*, 477–488. [CrossRef]
27. Costas-Chavarri, A.; Nandakumar, G.; Temin, S.; Lopes, G.; Cervantes, A.; Cruz Correa, M.; Engineer, R.; Hamashima, C.; Ho, G.F.; Huitzil, F.D.; et al. Treatment of Patients With Early-Stage Colorectal Cancer: ASCO Resource-Stratified Guideline. *J. Glob. Oncol.* **2019**, *5*, 1–19. [CrossRef]

28. Kotake, K.; Honjo, S.; Sugihara, K.; Hashiguchi, Y.; Kato, T.; Kodaira, S.; Muto, T.; Koyama, Y. Number of Lymph Nodes Retrieved is an Important Determinant of Survival of Patients with Stage II and Stage III Colorectal Cancer. *Jpn. J. Clin. Oncol.* **2012**, *42*, 29–35. [CrossRef]
29. Chang, G.J.; Rodriguez-Bigas, M.A.; Skibber, J.M.; Moyer, V.A. Lymph node evaluation and survival after curative resection of colon cancer: Systematic review. *J. Natl. Cancer Inst.* **2007**, *99*, 433–441. [CrossRef]
30. Compton, C.; Fenoglio-Preiser, C.M.; Pettigrew, N.; Fielding, L.P. American Joint Committee on Cancer Prognostic Factors Consensus Conference: Colorectal Working Group. *Cancer* **2000**, *88*, 1739–1757. [CrossRef]
31. Compton, C.C.; Greene, F.L. The staging of colorectal cancer: 2004 and beyond. *CA A Cancer J. Clin.* **2004**, *54*, 295–308. [CrossRef]
32. Bilimoria, K.Y.; Bentrem, D.J.; Stewart, A.K.; Talamonti, M.S.; Winchester, D.P.; Russell, T.R.; Ko, C.Y. Lymph node evaluation as a colon cancer quality measure: A national hospital report card. *J. Natl. Cancer Inst.* **2008**, *100*, 1310–1317. [CrossRef] [PubMed]
33. Babcock, B.D.; Aljehani, M.A.; Jabo, B.; Choi, A.H.; Morgan, J.W.; Selleck, M.J.; Luca, F.; Raskin, E.; Reeves, M.E.; Garberoglio, C.A.; et al. High-Risk Stage II Colon Cancer: Not All Risks Are Created Equal. *Ann. Surg. Oncol.* **2018**, *25*, 1980–1985. [CrossRef] [PubMed]
34. Gill, S.; Loprinzi, C.L.; Sargent, D.J.; Thomé, S.D.; Alberts, S.R.; Haller, D.G.; Benedetti, J.; Francini, G.; Shepherd, L.E.; Francois Seitz, J.; et al. Pooled analysis of fluorouracil-based adjuvant therapy for stage II and III colon cancer: Who benefits and by how much? *J. Clin. Oncol.* **2004**, *22*, 1797–1806. [CrossRef] [PubMed]
35. Benson, A.B.; Venook, A.P.; Al-Hawary, M.M.; Cederquist, L.; Chen, Y.-J.; Ciombor, K.K.; Cohen, S.; Cooper, H.S.; Deming, D.; Engstrom, P.F.; et al. NCCN Guidelines Insights: Colon Cancer, Version 2.2018. *J. Natl. Compr. Cancer Netw.* **2018**, *16*, 359–369. [CrossRef]
36. Lin, L.Y.; Warren-Gash, C.; Smeeth, L.; Chen, P.C. Data resource profile: The National Health Insurance Research Database (NHIRD). *Epidemiol. Health* **2018**, *40*, e2018062. [CrossRef]
37. Tsai, F.-J.; Junod, V. Medical research using governments' health claims databases: With or without patients' consent? *J. Public Health* **2018**, *40*, 871–877. [CrossRef]
38. Hsing, A.W.; Ioannidis, J.P. Nationwide Population Science: Lessons from the Taiwan National Health Insurance Research Database. *JAMA Intern. Med.* **2015**, *175*, 1527–1529. [CrossRef]

medicina

MDPI

Article

A Reduction of Calcineurin Inhibitors May Improve Survival in Patients with De Novo Colorectal Cancer after Liver Transplantation

Ramin Raul Ossami Saidy [1,*], Elisa Wegener [1], Deniz Uluk [1], Luca Dittrich [1], Wenzel Schöning [1], Georg Lurje [1], Robert Öllinger [1], Dominik Paul Modest [2], Frank Tacke [3], Oliver Haase [1], Johann Pratschke [1] and Dennis Eurich [1]

1. Department of Surgery, Campus Charité Mitte | Campus Virchow Klinikum, Charité-Universitätsmedizin Berlin, Charité University Medicine Berlin, 13353 Berlin, Germany
2. Department of Hematology, Oncology and Cancer Immunology, Campus Charité Mitte | Campus Virchow Klinikum, Charité-Universitätsmedizin Berlin, Charité University Medicine Berlin, 13353 Berlin, Germany
3. Department of Hepatology and Gastroenterology, Campus Charité Mitte | Campus Virchow Klinikum, Charité-Universitätsmedizin Berlin, Charité University Medicine Berlin, 13353 Berlin, Germany

* Correspondence: ramin-raul.ossami-saidy@charite.de; Tel.: +49-30-450-652557

Abstract: *Background and Objectives:* After liver transplantation (LT), long-term immunosuppression (IS) is essential. IS is associated with de novo malignancies, and the incidence of colorectal cancer (CRC) is increased in LT patients. We assessed course of disease in patients with de novo CRC after LT with focus of IS and impact on survival in a retrospective, single-center study. *Materials and Methods*: All patients diagnosed with CRC after LT between 1988 and 2019 were included. The management of IS regimen following diagnosis and the oncological treatment approach were analyzed: Kaplan–Meier analysis as well as univariate and multivariate analysis were performed. *Results*: A total of 33 out of 2744 patients were diagnosed with CRC after LT. Two groups were identified: patients with restrictive IS management undergoing dose reduction (RIM group, $n = 20$) and those with unaltered regimen (maintenance group, $n = 13$). The groups did not differ in clinical and oncological characteristics. Statistically significant improved survival was found in Kaplan–Meier analysis for patients in the RIM group with 83.46 (8.4–193.1) months in RIM and 24.8 (0.5–298.9) months in the maintenance group (log rank = 0.02) and showed a trend in multivariate cox regression ($p = 0.054$, HR = 14.3, CI = 0.96–213.67). *Conclusions*: Immunosuppressive therapy should be reduced further in patients suffering from CRC after LT in an individualized manner to enable optimal oncological therapy and enable improved survival.

Keywords: liver transplantation; de novo malignancy; colorectal carcinoma; immunosuppression

Citation: Saidy, R.R.O.; Wegener, E.; Uluk, D.; Dittrich, L.; Schöning, W.; Lurje, G.; Öllinger, R.; Modest, D.P.; Tacke, F.; Haase, O.; et al. A Reduction of Calcineurin Inhibitors May Improve Survival in Patients with De Novo Colorectal Cancer after Liver Transplantation. *Medicina* **2022**, *58*, 1755. https://doi.org/10.3390/medicina58121755

Academic Editors: Antonio M Scanu and Maria Rosaria De Miglio

Received: 29 October 2022
Accepted: 28 November 2022
Published: 29 November 2022

1. Introduction

Liver transplantation (LT) is still the only option for various conditions resulting in end-stage liver disease as well as primary malignancies of the liver itself. After LT, life-long or at least long-term immunosuppression (IS) remains standard for the prevention of graft rejection. Here, calcineurin inhibitors (CNI), mycophenolate mofetil (MMF), glucocorticoids (GC) and mammalian target of rapamycin inhibitors (mTORI) are most frequently used, and their fine-tuned regimen is one of the main reasons for the markedly prolonged survival of graft function after LT in the last decades [1]. However, side effects such as chronic kidney injury and neoplasms in the decade-long administration of CNI are well known, and the overall beneficial effects of mTORI are controversial [2,3].

With increasing graft survival, long-term outcomes after LT including comorbidities and complications of IS therapy are gaining more interest. For example, the risk of de novo malignancies (DNM) in patients after LT is significantly elevated with an reported incidence

2- to 3-fold compared with the general population [4,5]. Further, cancer-associated mortality is expected to become the most frequent cause of death in the cohort of LT patients and is already the leading cause of death in the second decade after transplantation [6–8].

Colorectal cancer (CRC) is one of the most common malignancies worldwide, and its incidence is elevated after LT [9,10]. The stage-dependent therapeutic regimen is highly standardized and consists of radiotherapy, chemotherapy, surgical resection and optional antibody treatment regarding the individual profile. Compared with the overall population, CRC in LT patients is associated with an increased incidence, comparable with the overall rate of DNMs, and occurrence is reported to be earlier in life [11,12]. Of note, certain underlying diseases leading to LT such as PSC alone or in coincidence with inflammatory bowel diseases (IBD) elevate the risk of development of CRC even further to more than seven times [13,14]. Additionally, non-alcoholic liver disease and hepatocellular carcinoma (HCC) have been associated with increased risk after LT [15]. Reports of outcome after CRC in LT patients are heterogenous. Comparable survival rates have been shown, but also poorer long-term survival in patients after solid organ transplantation [11,16]. However, the handling and especially the clinical impact of modification of IS for LT after the diagnosis of de novo CRC remain unclear, and scientific data are not available, although recommendations have been established recently [17,18].

Previously, we investigated the effect of reduction of immunosuppression in patients suffering from recurrent primary liver malignancies such as hepatocellular carcinoma (HCC) or lung cancer after LT and found an impact on survival for patients with dose reduction upon diagnosis independent of oncological treatment [19,20]. In this study, we investigate patients' course after diagnosis of de novo CRC after LT with a focus on the impact of immunosuppressive management.

2. Patients and Methods

Patients undergoing LT for various conditions at our institution between 1988 and 2020 and with diagnosis of de novo CRC post LT were included in the analysis. Diagnosis of CRC was confirmed by histopathology, and staging was conducted according to guidelines using the classification of the Union for International Cancer Control (UICC) based upon the TNM-classification [21,22]. Oncological regimen was categorized into curative or palliative and best supportive care (BSC).

After LT, all patients were followed up periodically at our outpatient center. Intervals were based on the time after transplantation, ranging from two times a week to every twelve weeks. Here, clinical and laboratory examinations were conducted, and ultrasound-guided, transcostal needle biopsies of the graft were performed according to internal standard protocol at 1, 3, 5, 7, 10 and 13 years and on individual basis thereafter. Routine surveillance via colonoscopy was conducted as recommended by current guidelines, but with intervals of at least five years and intensified surveillance in patients suffering from inflammatory bowel disease (IBD) ranging from once or twice per year to individual intervals as recommended by treating endoscopists [23,24].

To evaluate IS, a score first introduced by Vasudev et al. was used, allowing semiquantitative comparability of different substances (one unit for each daily dose of: prednisone—5 mg, cyclosporine a—100 mg, tacrolimus—2 mg, MMF—500 mg, sirolimus—2 mg) [25]. Cumulative Vasudev score calculated by addition of score over the years and median score were evaluated. Using the approach presented by Rodríguez-Perálvarez et al., impacts of tacrolimus trough levels were analyzed after classification into minimized exposure (<5 ng/mL) and conventional exposure (>5 ng/mL) [26]. Here, mean trough level was calculated (at least one measurement/year) after diagnosis of CRC. For assessment of impact of IS after diagnosis of CRC, management of immune suppressive regimen was grouped in two categories for analysis: (i) maintaining immunosuppression or (ii) new restrictive immunosuppressive management (RIM). RIM was defined when dose reduction or complete discontinuation of IS after diagnosis of cancer was documented. Of note, alteration of mTOR therapy was classified differently: initiation of mTORI without reduction of prior IS was classified as (i) and only if concomitant reduction of other IS (CNI, GC, MMF) was performed were these cases grouped in (ii). Oncological course of patients was followed up by in-hospital data and reports from corresponding institutions, as therapy for LT patients was outlined in an interdisciplinary approach with primary care physicians and oncologists. Thus, data on clinical course as well as laboratory, histological or radiological parameters were extracted from our prospectively maintained database.

Statistical analysis was performed using SPSS Statistics Version 26.0 (IBM Co., Armonk, NY, USA). By its retrospective character, the study design was exploratory. For the testing of statistically significant differences, cross-tables were used for nominal-scaled variables. T-test was applied for continuous, normal-distributed variables. For the testing of non-normally distributed values, the Mann–Whitney U-test or Kruskal–Wallis test were chosen. For the analysis of impact on survival, univariate analysis and Kaplan–Meier analysis were conducted, and log rank tests were calculated. To evaluate effect strength, multivariate and univariate Cox regression models were used, and hazard ratio (HR) and confidence interval (CI) were calculated. Putative relevant variables or confounders for integration in multivariate analysis were identified by clinical experience, such as patients´ characteristics (relevant comorbidities, age, sex) or oncological parameters. A p-value of <0.05 was considered significant.

The study was conducted in accordance with the guidelines of the Declaration of Helsinki and was approved by the local ethics committee of our institution (protocol code EA1/255/20; date of approval: 20 October 2020).

3. Results

From 2744 patients receiving LT over a 33-year span, 33 patients were identified with de novo colorectal cancer, forming a prevalence of 1.2% in this population. Median time from transplantation to DNM was 12.0 years (0.9–27). Indications for initial LT and overall patient characteristics are displayed in Table 1. Prior to the diagnosis of CRC, immunosuppressants used were CNI (n = 28; 84.8%), MMF (n = 7; 21.2%), mTORI (n = 4; 12.1%) and glucocorticoids (n = 1; 3%). A group of 31 (93.9%) patients were diagnosed with colon cancer and two (6.1%) with rectal cancer. Using the UICC criteria, 14 (42.4%) patients were stage I, eight (24.2%) stage II, six (18.2%) stage III and three (9.1%) stage IV at initial diagnosis. Based on staging and patients' constitution, 32 (97.0%) patients were treated with curative and only one (3.0%) patient with palliative intention. Regimens consisted of oncological resection in 32 (97.0%) cases, chemotherapy in nine (27.3%) and radiotherapy in two (6.1%), with either combination in eight (24.4%) cases based on therapy standards at the specific time. Adjuvant chemotherapy was administered in eight (24.2%) cases, and only one patient received palliative chemotherapy (3.0%). Median survival after diagnosis of de novo CRC was 49.6 (0.5–298.9) months. At the end time of observation, 11 (33.3%) patients had died, and in eight (24.2%), the malignancy was stated as cause of death. In all patients undergoing surgery, histopathology confirmed local R0-resection. We did not find statistical impact of T-stadium or N-classification on survival in Kaplan–Meier analysis, but M1 status was associated with significant shorter survival (log rank 0.001). Kaplan–Meier analysis also revealed the statistical significance of UICC stage on survival after diagnosis with a median of 66.1 (2–129.2) months in stage I, 88.8 (8.4–298.9) months in stage II, 48.8 (0.5–193.1) months in stage III and 36.8 (3.2–55.4) months in stage IV (log rank < 0.01). Regarding decade of diagnosis (1989–1999/2000–2009/2010–2019/2020-today), to account for different oncological therapeutic options, no impact on overall survival after diagnosis was found (log rank = 0.52).

Table 1. Overall cohort and Group characteristics.

	All Patients (n = 33)	RIM (n = 20)	no RIM (n = 13)	p *
Median age at LT years (min-max)	51.0 (14.0–61.0)	50.0 (29–59)	51.0 (14–61)	0.75
sex (%)				
male	19 (57.6)	11 (55)	8 (61.5)	0.5
female	14 (42.2)	9 (45)	5 (38.5)	
Indication for liver transplantation (%)				
ALD	8 (24.2)	5 (25.0)	3 (23.1)	
PBC/PSC	9 (27.3)	6 (30.0)	3 (23.1)	
HCC/CCC	4 (12.1)	4 (20.0)	0 (0)	0.35
viral hepatitis	5 (15. 2)	3 (15.0)	2 (15.4)	
others	7 (21.2)	2 (10.0)	5 (38.4)	
Induction of immunosuppression (%)				
none	11 (33.3)	6 (30.0)	5 (38.5)	
antibodies	16 (48.5)	12 (60.0)	4 (30.8)	0.18
ATG	6 (18.2)	2 (10.0)	4 (30.8)	
Immunosuppression at diagnosis of CRC (%)				
CNI	28 (84.8)	19 (95.0)	9 (69.2)	
MMF	7 (21.2)	4 (20.0)	3 (23.1)	
GC	4 (12.1)	2 (10.0)	2 (15.4)	0.61
mTORI	5 (15.2)	3 (15.0)	2 (15.4)	
combination	6 (18.2)	5 (25.0)	1 (7.7)	
Cardiovascular comorbidities at diagnosis of CRC	24 (72.3)	16 (80.0)	8 (66.7)	0.43
IBD	7 (21.2)	5 (25.0)	2 (15.4)	0.68
BMI at diagnosis of CRC kg/m^2 (min-max)	23.0 (16–36)	24.9 (16–36)	22.6 (18–27)	0.12
Median age at diagnosis of CRC years (min-max)	60.0 (29–79)	63.0 (36–78)	60.0 (29–79)	0.59
Median time to CRC after LT years (min-max)	12.0 (0.9–27)	12.5 (1.0–29.0)	11.0 (0.9–27.0)	0.44
Decade at time of CRC (%)				
1989–1990	4 (12.1)	1 (5.0)	3 (23.1)	
2000–2009	9 (27.3)	7 (35.0)	2 (15.4)	0.34
2010–2019	18 (54.5)	11 (55.0)	7 (53.8)	
2020-today	2 (6.1)	1 (5.0)	1 (7.7)	
UICC stage				
I	14 (42.4)	10 (50.0)	4 (36.4)	
II	8 (24.2)	4 (20.0)	4 (36.4)	
III	6 (18.2)	5 (20.0)	1 (9.1)	0.36
IV	3 (9.1)	1 (5.0.)	2 (18.2)	
missing	2 (6.1)	-	2 (15.4)	
curative oncological regimen	32 (97.0)	20 (100.0)	12 (92.3)	0.34
Deceased at follow-up	11 (33.3)	4 (20)	7 (53.8)	
cause of death				
CRC	7 (21.2)	2 (10)	5 (38.5)	0.38
cardiovascular	3 (9.1)	1 (5)	2 (15.4)	
other	1 (3.0)	1 (5)	0	

LT—liver transplantation; ALD—alcoholic liver disease; PBC—primary biliary cholangitis; PSC—primary sclerosing cholangitis; HCC-hepatocellular carcinoma; CCC—cholangiocellular carcinoma; ATG—anti-thymocyte globuline; CRC—colorectal cancer; IBD—inflammatory bowel disease; CNI—calcineurine inhibitor; MMF—mycophenolate mofetile; GC—glucocorticoid; mTORI—mammalian target of rapamycin inhibitor; BMI—body mass index; UICC—Union for International Cancer Control. *—comparison of RIM and no RIM.

Median IS-score assessed according to Vasudev et al. at time of diagnosis was 2.0 (0.25–6.0) units, and median cumulative IS-score was 30.5 (3.0–87.5). After diagnosis of CRC, 20 (60.6%) patients were identified, where reduction of immunosuppression according to RIM-criteria in response to new malignancy was initiated. Thus, two groups were formed termed RIM and maintenance, respectively. In four patients, IS was withdrawn completely. Mean IS-score did not differ between groups at time of diagnosis with 2.1 (±1.5) units in group RIM and 2.5 (±1.4) units in maintenance group (p = 0.5). In RIM-patients, reduction of CNI was initiated in all patients, with relative dosage reduction of 45.0% (0.25–1). Additionally, MMF was reduced in four (20.0%) patients. In four (20.0%) patients, mTORI was introduced into regimen. Immune suppressive regimen prior to the diagnosis of CRC did not differ between the two groups with CNIs as backbone in 19 (95.6%) patients in RIM and in nine (69.3%) patients in the other group. The Wilcoxon test for non-parametric paired variables revealed a dose reduction of IS with statistical significance with an IS-score after prior to diagnosis of 2.1 (±1.5) units and 1.4 (±1.5) after diagnosis of CRC in the RIM-group ($p < 0.01$).

The most frequent indications for LT were alcoholic liver disease (ALD) and primary biliary cholangitis (PBC)/primary sclerosing cholangitis (PSC) in both groups without significant differences (p = 0.35). Further, the prevalence of inflammatory bowel disease (IBD) did not differ between groups (p = 0.68). Median time to de novo CRC was comparable (RIM: 12.5 (1.0–29.0) years/maintenance: 11.0 (0.9–27.0) years, p = 0.44). Furthermore, stage of malignancy using the UICC classification showed no significant difference between groups; most patients were diagnosed with local tumor stages of I/II in 14 (70.0%) patients in the group with restrictive IS management and eight (72.8%) in those with unaltered IS-regimen (p = 0.36). Table 1 shows an overview of patient characteristics including oncological parameters. Here, no statistically significant differences between those two groups were found. Additionally, no rejection or loss of graft occurred in the group undergoing further reduction of IS, and thus, no patient received a re-installment of a previous IS-regimen.

Median survival from initial diagnosis was 83.46 (8.4–193.1) months in the RIM group and 24.8 (0.5–298.9) months in maintenance. At the end of the observation period, four patients (20.0%) had died under restrictive immunosuppression and seven (46.2%) in the group of unaltered IS. Cause of death was CRC in two (20.0%) and five (38.5%). No significance was found in causes of death between groups (p = 0.38; see Table 1). Comparison using Kaplan–Meier survival analysis showed statistically significant differences in both short-term and long-term survival (log rank = 0.02); see also Figure 1.

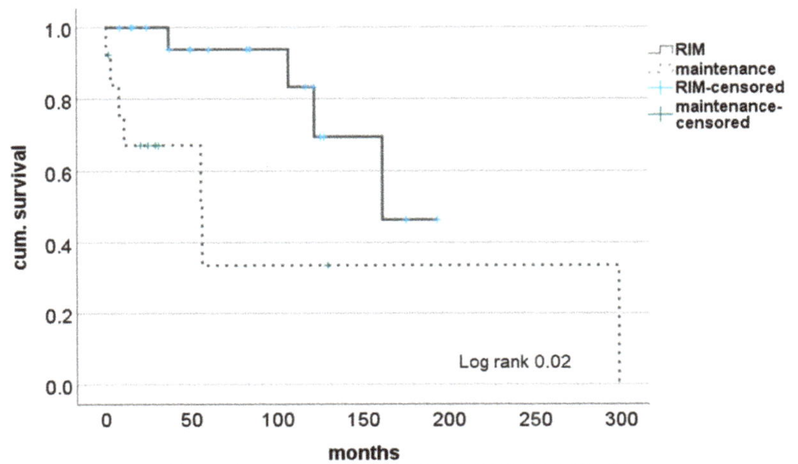

N° at risk							
RIM	20	12	9	3	0	0	0
maintenance	13	4	2	1	1	1	1

Figure 1. Impact of RIM on survival after diagnosis of de novo CRC after LT. RIM—restrictive immunosuppressive management.

We did not find improved survival after the diagnosis of CRC for the five (15.2%) patients receiving mTORI before compared with those without (log rank 0.13) or those five (15.2%) with mTORI therapy after diagnosis (log rank 0.29).

The subgroup analysis of patients with regard to N- and M-status showed trends for a survival benefit for patients with RIM but did not reach statistical significance except for short-term survival in patients with M1 status (see Figure 2). Analyzing the survival of patients with or without RIM subgrouped for UICC stage showed no impact in stages I and II but significantly longer survival for patients with UICC stages III and IV when a restrictive immune suppressive regimen after diagnosis of CRC was conducted. Here, median survival was 48.8 (16.2–193.1) months and 3.2 (0.5–55.4) months, respectively (log rank 0.02); see Figure 3.

(a)

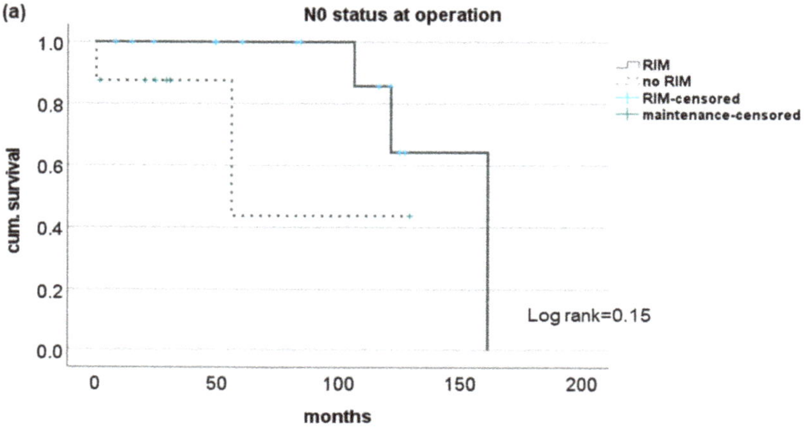

N° at risk					
RIM	14	10	7	1	0
maintenance	6	2	1	0	0

(b)

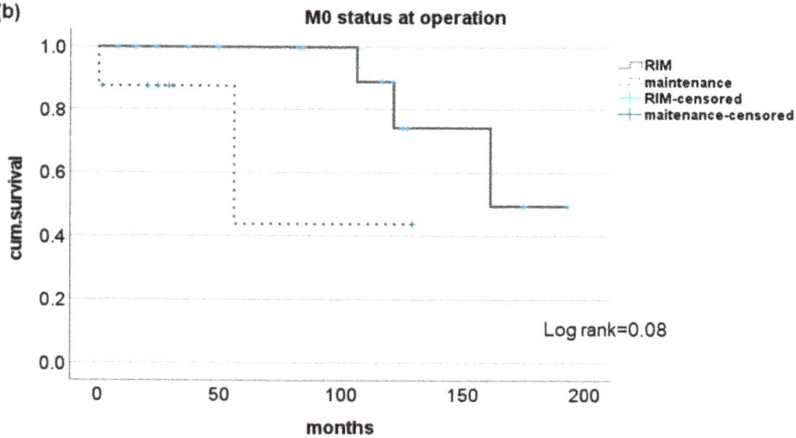

N° at risk					
RIM	17	11	9	3	0
maintenance	8	2	1	0	0

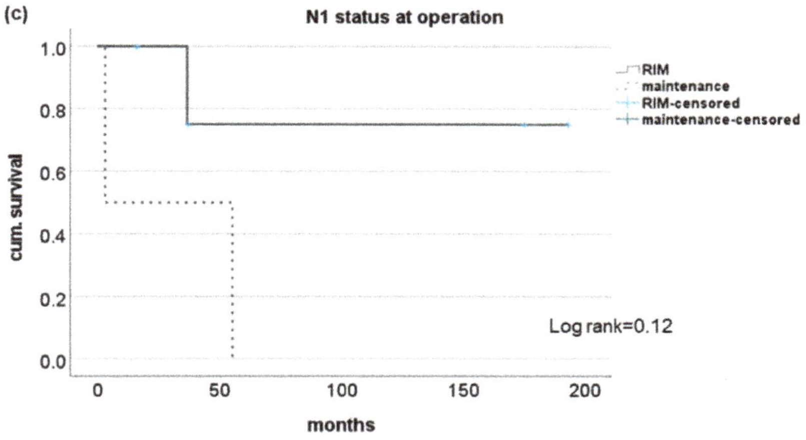

N° at risk					
RIM	4	2	1	1	0
maintenance	1	0	0	0	0

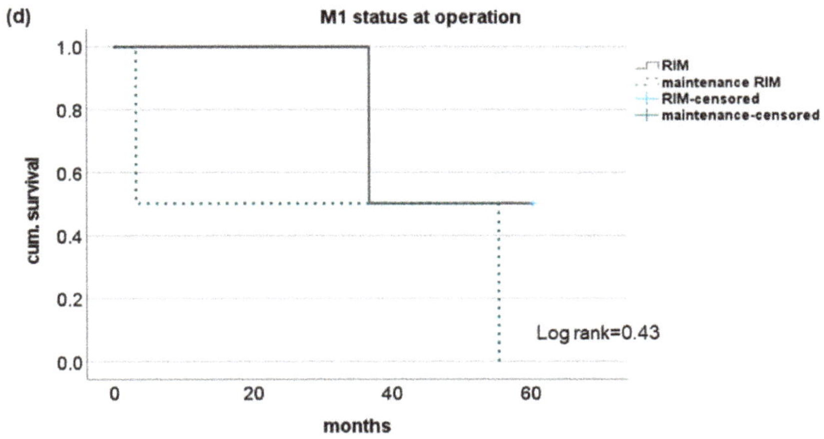

N° at risk				
RIM	2	1	0	0
maintenance	2	0	0	0

Figure 2. Subgroup analysis of survival of patients with and without RIM dependent on lymph node manifestation or distant metastases at time of diagnosis of CRC after LT. Kaplan-Meier analysis of patients without tumor manifestations in lymph nodes (**a**) or distant metastases (**b**) as well as patients with histological proven tumor manifestation in local lymph nodes (**c**) or distant metastases (**d**) at initial diagnosis of CRC seem to profit from a additional restrictive immunosuppressive regimen upon diagnosis but no statistical significant difference was reached. RIM—restrictive immunosuppressive management.

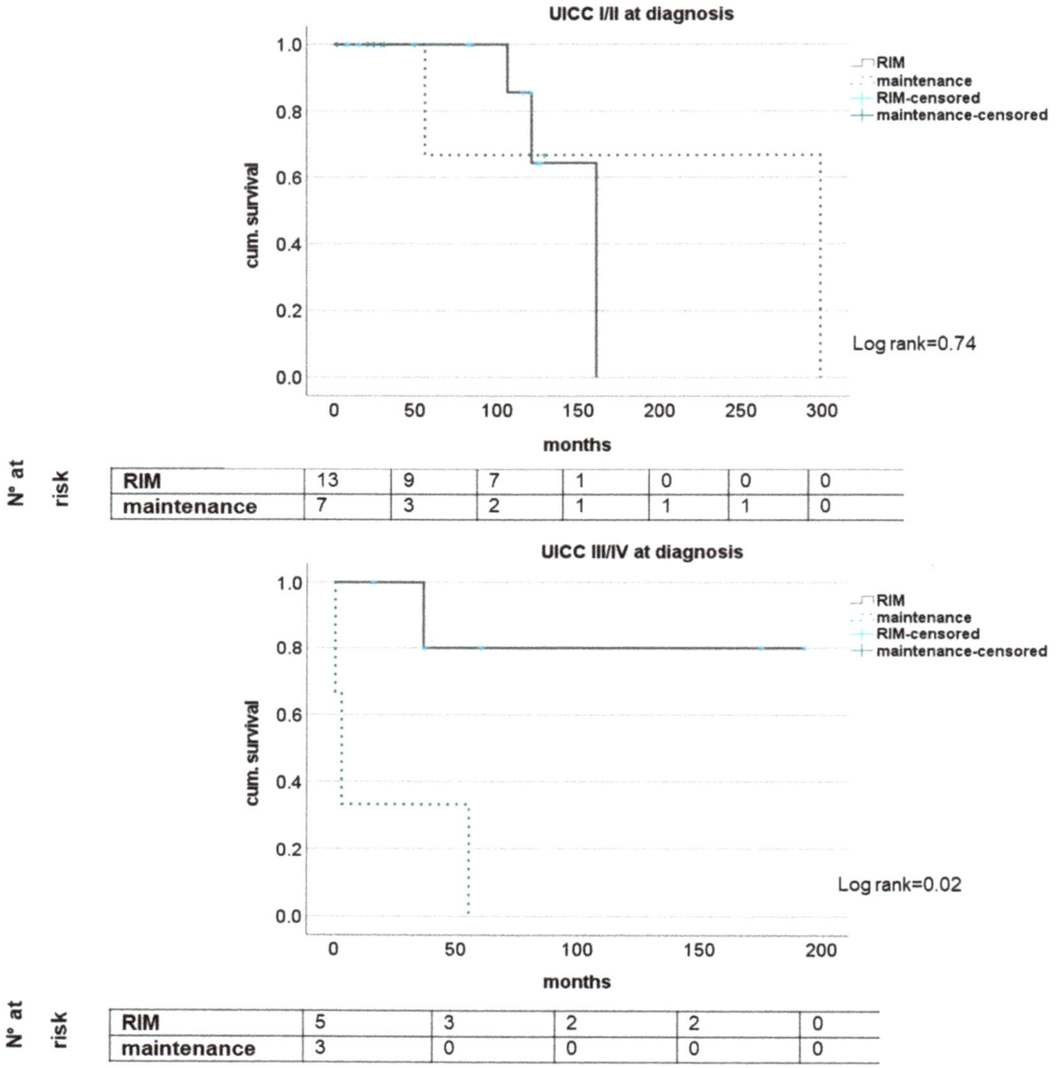

Figure 3. Impact of RIM dependent on tumor stage according to UICC.

In multivariate analysis using the clinically important variables of age at tumor diagnosis and preexistent cardiovascular disease and the oncological staging parameters using the TNM classification and RIM, no significant statistical impact on improved overall survival after diagnosis of de novo CRC after LT was found. However, a trend regarding impact of RIM was seen ($p = 0.054$); see also Table 2.

Analyzing the impact of tacrolimus trough levels after diagnosis of CRC, we found significantly improved survival in groups with mean trough levels of <5 ng/mL (minimized exposure) and >5 ng/mL (conventional exposure) after diagnosis of de novo CRC, (log rank 0.03); see Figure 4.

Table 2. Multivariate Regression analysis on impact of survival after diagnosis of CRC after LT.

Parameter	*p*	Hazard Ratio	95% CI	
			Lower	Upper
age at diagnosis of CRC ($\leq$55 vs. >55 years)	0.830	0.708	0.030	16.511
cardiovascular disease (reference: yes)	0.820	1.277	0.156	10.458
T (reference: T1)	0.250	2.349	0.548	10.062
N (reference: N1)	0.354	0.375	0.047	2.982
M (reference: M1)	0.102	6.439	0.690	60.112
RIM (reference: no RIM)	0.054	14.321	0.960	213.671

CRC—colorectal cancer; RIM—restrictive immunosuppressive management.

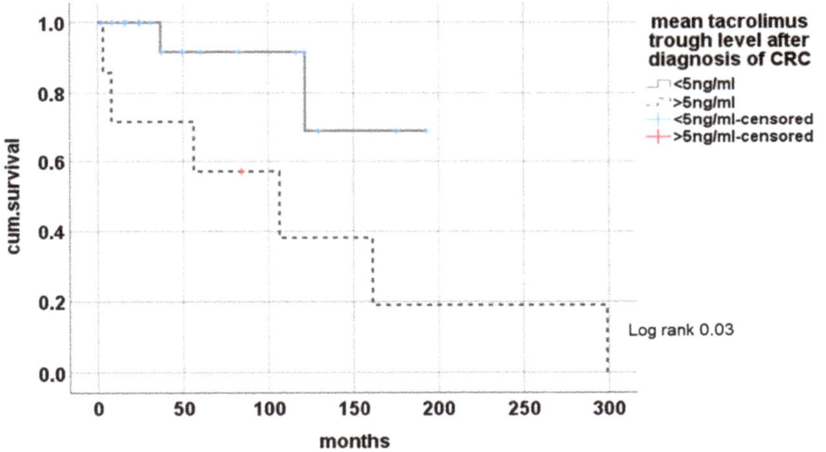

N° at risk							
RIM	18	8	6	2	0	0	0
maintenance	7	5	3	1	0	1	0

Figure 4. Dependence of survival from tacrolimus trough level after de novo CRC.

4. Discussion

In this study, we analyzed the course of patients with CRC after LT. Focus was current IS and its impact on survival, as its influence gains more relevance in recent studies on outcome after LT with regard on long-term survival [26–28].

We only found 33 patients out of our cohort of over 2700 patients in a time span of three decades with reported manifestation of CRC, forming a total prevalence of 1.2%, highlighting effective colorectal cancer screening. Studies report an incidence in the general population between 30 and 50/100,000 of new CRC per year in western countries, and an incidence of CRC in LT patients of 4.9% was reported by Altieri et al. [29–31]. Due to the life-long follow-up of our patients with high compliance, we do not expect underreporting in our collective but excellent patient adherence to our recommended follow-up examinations that include endoscopies after LT, and thus, many precancerous lesions might have been treated before the manifestation of CRC. This notion was supported by the high fraction of UICC stage I and stage II CRC that formed two thirds of our cohort. Subsequently, curative surgical therapy was available in every patient but one. We found a median occurrence of CRC after LT of 12 years, reflecting the impact of chronic IS and the shift in comorbidities that challenge the aftercare of patients after LT in the long run. Staging-dependent survival rates in our study were comparable with the general population and with LT patients from other reports [13,32]. Staging using the UICC criteria for CRC and the TNM classification demonstrated prognostic value in our cohort, reflecting their importance in decision making [33–36]. As most patients (all but one) were treated with curative intention with surgical resection, the most relevant impact for survival—surgical resectability—could not be assessed in our study, thus, however, an important potential bias for our study was ruled out in favor of the impact of IS-redesign.

Evaluating the effect of altered handling of IS after diagnosis of CRC, we found the two groups that were formed comparable in all relevant clinical aspects. Thus, impact of RIM could be assessed with validity. Survival analysis revealed positive effects of reducing IS further after de novo malignancy in LT patients, similar to findings for patients suffering from recurrent HCC after LT and in congruence of pathophysiology of administered substances [19,37]. The effect did not reach statistical significance in multivariate analysis, possibly to the very small population. Analyzing the effect of RIM in subgroups, we found impact especially in stages where tumor manifestation was advanced (UICC stages III/IV, M1-status at time of diagnosis). While the utmost importance with highest impact lies undoubtedly within stage-dependent oncological regimen, we hypothesize that the effect of RIM might become evident in cases where overall systemic immune control is overwhelmed, reflecting advanced stages [38–40]. As most patients suffering from CRC after LT are found years after the initial transplantation with stable liver function—as indicated in our cohort by the feasibility of major visceral operation—we conclude that RIM should be evaluated as an additional oncological aspect in this special cohort of patients with the aim of complete withdrawal. While early withdrawal has been shown to be of only minor success in certain subsets of patients, long-term discontinuation seems to be more favorable and feasible [41–44]. However, in an event of a life-threatening disease associated with failed immune response, we deem it mandatory to investigate its practicability in every individual in a step-by-step manner [45,46]. Recently, Colmenero et al. presented guidelines from the ILTS-SETH Consensus Conference regarding the incidence and management of DNMs [17]. While they note the lack of data altogether and the practical absence of prospective studies, their recommendations reflect this study´s findings.

The exact approach to reducing IS in LT patients remains partly unclear and always requires knowledge of the individual patient´s risk profile, comorbidities and tolerance to different substances and their adverse effects [47–49]. CNIs remain the most important substance, and all patients undergoing RIM in our study were found with reductions in this drug class. Additionally, we found a tacrolimus through-level-dependent survival difference with beneficial outcome for patients with lower CNI burden. In contrast, mTORI are the only substance of IS where anti-proliferative properties are reported, although its clinical impact remains controversial, and optimal regimen is unclear [50–56]. We did not find any impact of mTORI on survival, whether administered before or after the diagnosis of CRC, but the number of patients with mTORI was very low. In this regard, using the IS scale proposed by Vasudev et al. might be misguided, as mTORI are weighted equally to CNI, and from our regard, the influence of MMF might be overestimated [25]. However, using the IS scale, a low immunosuppressive burden in the overall cohort was shown, reflecting the modern approach of reducing IS after LT to the tolerable minimum.

Certain limitations of this study have to be addressed. The retrospective, three-decade-spanning character certainly inherited different approaches in post-LT management as well as oncological strategies and therapeutic options that were not explored in depth. Additionally, while the low number of patients reflects the rarity of this special constellation, it especially limited validity for subgroup analysis and also restricts overall statistical analysis. The use of different immune suppressants in over 30 years of LT with diverging focuses (preventing rejection at all cost vs. minimizing adverse side effects for the future) is certainly present in this study and the calculation of IS-score and definition of RIM may be unprecise. However, strategies for CRC have made enormous progress, and regimens including total neoadjuvant concepts for rectal cancer and targeted therapies for metastatic conditions as well as extended concepts for colorectal liver metastases have improved survival for patients immensely [57–61]. We did not evaluate the differentiated oncological strategies, but the distribution of diagnosis of CRC over the decades did not differ between our groups, and thus, possible bias of diverging options even for advanced stages should be ruled out. It has to be acknowledged that while this study further confirms recent recommendations, it inherits the methodical limitations of the few studies published before investigating this issue. Thus, the presented collective can only be regarded as an addition to the growing, but still scarcely existing, body of evidence.

5. Conclusions

A remarkable oncological benefit for a restrictive, reflective management of IS upon diagnosis of CRC after LT with significant impact on survival for the individual patients was found in this study. This observation requests timely action from the physician in charge after LT in an individualized manner with close correspondence to treating oncologists as IS reduction can be regarded as an additional oncological measure. To achieve a profound scientific foundation for the reduction of IS in this context, prospective, multi-center data must be acquired in regard to the rarity of occurrence.

Author Contributions: Conceptualization, R.R.O.S., E.W. and D.E.; methodology, E.W., D.U., D.E. and R.R.O.S.; software, R.R.O.S., L.D.; validation, W.S., G.L., R.Ö. and F.T.; formal analysis, R.R.O.S. and D.E.; investigation, E.W. and R.R.O.S.; resources, J.P.; data curation, R.R.O.S. and D.U.; writing—original draft preparation, R.R.O.S., D.E., D.P.M. and O.H.; writing—review and editing, R.R.O.S., D.E., O.H. and F.T.; visualization, R.R.O.S. and L.D.; supervision, J.P., D.P.M., F.T. and O.H.; project administration, D.E.; All authors have read and agreed to the published version of the manuscript.

Funding: There was no financial support or funding received by the authors related to the presented work.

Institutional Review Board Statement: The study was conducted in accordance with the guidelines of the Declaration of Helsinki and was approved by the local ethics committee of our institution (protocol code EA1/255/20; date of approval: 20 October 2020).

Informed Consent Statement: Informed consent was obtained from all subjects involved in the study.

Data Availability Statement: The data presented in this study are available on request from the corresponding author. The data are not publicly available due to local ethics committee stipulations and privacy policy of Charité-Universitätsmedizin Berlin.

Conflicts of Interest: All authors declare no conflicts of interest related to the presented work.

Abbreviations

ALD	alcoholic liver disease
BSC	best supportive care
CCA	cholangiocellular carcinoma
CI	confidence interval
CNI	calcineurininhibitor
CRC	colorectal carcinoma
DNM	de novo malignancy
GC	glucocorticoids
HCC	hepatocellular carcinoma
HR	hazard ratio
IBD	inflammatory bowel disease
IS	immunosuppression
LT	liver transplantation
MMF	mycophenolate mofetil
mTORI	mammalian target of rapamycin inhibitor
RIM	restrictive immunosuppressive management

PBC	primary biliary cholangitis
PSC	primary sclerosing cholangitis
UICC	Union for International Cancer Control

References

1. Di Maira, T.; Little, E.C.; Berenguer, M. Immunosuppression in liver transplant. *Best Pract. Res. Clin. Gastroenterol.* **2020**, *46–47*, 101681. [CrossRef] [PubMed]
2. Geissler, E.K.; Schnitzbauer, A.A.; Zülke, C.; Lamby, P.E.; Proneth, A.; Duvoux, C.; Burra, P.; Jauch, K.W.; Rentsch, M.; Ganten, T.M.; et al. Sirolimus Use in Liver Transplant Recipients With Hepatocellular Carcinoma: A Randomized, Multicenter, Open-Label Phase 3 Trial. *Transplantation* **2016**, *100*, 116–125. [CrossRef] [PubMed]
3. Klintmalm, G.B.; Nashan, B. The Role of mTOR Inhibitors in Liver Transplantation: Reviewing the Evidence. *J. Transplant.* **2014**, *2014*, 845438. [CrossRef] [PubMed]
4. Collett, D.; Mumford, L.; Banner, N.R.; Neuberger, J.; Watson, C. Comparison of the incidence of malignancy in recipients of different types of organ: A UK Registry audit. *Am. J. Transplant.* **2010**, *10*, 1889–1896. [CrossRef] [PubMed]
5. Engels, E.A.; Pfeiffer, R.M.; Fraumeni, J.F., Jr.; Kasiske, B.L.; Israni, A.K.; Snyder, J.J.; Wolfe, R.A.; Goodrich, N.P.; Bayakly, A.R.; Clarke, C.A.; et al. Spectrum of cancer risk among US solid organ transplant recipients. *JAMA* **2011**, *306*, 1891–1901. [CrossRef] [PubMed]
6. Chandok, N.; Watt, K.D. Burden of de novo malignancy in the liver transplant recipient. *Liver Transpl.* **2012**, *18*, 1277–1289. [CrossRef]
7. Idossa, D.W.; Simonetto, D.A. Infectious Complications and Malignancies Arising After Liver Transplantation. *Anesthesiol. Clin.* **2017**, *35*, 381–393. [CrossRef]
8. Daniel, K.E.; Eickhoff, J.; Lucey, M.R. Why do patients die after a liver transplantation? *Clin. Transplant.* **2017**, *31*, e12906. [CrossRef]
9. Burra, P.; Rodriguez-Castro, K.I. Neoplastic disease after liver transplantation: Focus on de novo neoplasms. *World J. Gastroenterol.* **2015**, *21*, 8753–8768. [CrossRef]
10. Mármol, I.; Sánchez-de-Diego, C.; Pradilla Dieste, A.; Cerrada, E.; Rodriguez Yoldi, M.J. Colorectal Carcinoma: A General Overview and Future Perspectives in Colorectal Cancer. *Int. J. Mol. Sci.* **2017**, *18*, 197. [CrossRef]
11. Kim, M.; Kim, C.W.; Hwang, S.; Kim, Y.H.; Lee, J.L.; Yoon, Y.S.; Park, I.J.; Lim, S.B.; Yu, C.S.; Kim, J.C.; et al. Characteristics and Prognosis of Colorectal Cancer after Liver or Kidney Transplantation. *World J. Surg.* **2021**, *45*, 3206–3213. [CrossRef] [PubMed]
12. Nicolaas, J.S.; De Jonge, V.; Steyerberg, E.W.; Kuipers, E.J.; Van Leerdam, M.E.; Veldhuyzen-van Zanten, S.J.O. Risk of colorectal carcinoma in post-liver transplant patients: A systematic review and meta-analysis. *Am. J. Transplant.* **2010**, *10*, 868–876. [CrossRef]
13. Rompianesi, G.; Ravikumar, R.; Jose, S.; Allison, M.; Athale, A.; Creamer, F.; Gunson, B.; Manas, D.; Monaco, A.; Mirza, D.; et al. Incidence and outcome of colorectal cancer in liver transplant recipients: A national, multicentre analysis on 8115 patients. *Liver Int.* **2019**, *39*, 353–360. [CrossRef] [PubMed]
14. Singh, S.; Edakkanambeth Varayil, J.; Loftus, E.V., Jr.; Talwalkar, J.A. Incidence of colorectal cancer after liver transplantation for primary sclerosing cholangitis: A systematic review and meta-analysis. *Liver Transpl.* **2013**, *19*, 1361–1369. [CrossRef] [PubMed]
15. Nasser-Ghodsi, N.; Mara, K.; Watt, K.D. De Novo Colorectal and Pancreatic Cancer in Liver-Transplant Recipients: Identifying the Higher-Risk Populations. *Hepatology* **2021**, *74*, 1003–1013. [CrossRef]
16. Merchea, A.; Shahjehan, F.; Croome, K.P.; Cochuyt, J.J.; Li, Z.; Colibaseanu, D.T.; Kasi, P.M. Colorectal Cancer Characteristics and Outcomes after Solid Organ Transplantation. *J. Oncol.* **2019**, *2019*, 5796108. [CrossRef]
17. Colmenero, J.; Tabrizian, P.; Bhangui, P.; Pinato, D.J.; Rodríguez-Perálvarez, M.L.; Sapisochin, G.; Bhoori, S.; Pascual, S.; Senzolo, M.; Al-Adra, D.; et al. De Novo Malignancy After Liver Transplantation: Risk Assessment, Prevention, and Management-Guidelines From the ILTS-SETH Consensus Conference. *Transplantation* **2022**, *106*, e30–e45. [CrossRef]
18. Hoppmann, N.; Massoud, O. Medical care of liver transplant patients. *Expert Rev. Gastroenterol. Hepatol.* **2020**, *14*, 901–918. [CrossRef]
19. Ossami Saidy, R.R.; Postel, M.P.; Pflüger, M.J.; Schoening, W.; Öllinger, R.; Gül-Klein, S.; Schmelzle, M.; Tacke, F.; Pratschke, J.; Eurich, D. Minimization of Immunosuppressive Therapy Is Associated with Improved Survival of Liver Transplant Patients with Recurrent Hepatocellular Carcinoma. *Cancers* **2021**, *13*, 1617. [CrossRef]
20. Pesthy, S.; Wegener, E.; Ossami Saidy, R.R.; Timmermann, L.; Uluk, D.; Aydin, M.; Dziodzio, T.; Schoening, W.; Lurje, G.; Öllinger, R.; et al. Reducing Immunosuppression in Patients with De Novo Lung Carcinoma after Liver Transplantation Could Significantly Prolong Survival. *Cancers* **2022**, *14*, 2748. [CrossRef]
21. Greene, F.L. Current TNM staging of colorectal cancer. *Lancet Oncol.* **2007**, *8*, 572–573. [CrossRef] [PubMed]
22. James, D.; Brierley, M.K.G. Christian Wittekind. In *TNM Classification of Malignant Tumours*, 8th ed.; Wiley and the UICC: Hoboken, NJ, USA, 2016; Volume 1.
23. Shaukat, A.; Kahi, C.J.; Burke, C.A.; Rabeneck, L.; Sauer, B.G.; Rex, D.K. ACG Clinical Guidelines: Colorectal Cancer Screening 2021. *Off. J. Am. Coll. Gastroenterol. ACG* **2021**, *116*, 458–479. [CrossRef] [PubMed]
24. Available online: https://register.awmf.org/de/leitlinien/detail/021-007OL (accessed on 20 October 2022).
25. Vasudev, B.; Hariharan, S.; Hussain, S.A.; Zhu, Y.R.; Bresnahan, B.A.; Cohen, E.P. BK virus nephritis: Risk factors, timing, and outcome in renal transplant recipients. *Kidney Int.* **2005**, *68*, 1834–1839. [CrossRef] [PubMed]

26. Rodríguez-Perálvarez, M.; Colmenero, J.; González, A.; Gastaca, M.; Curell, A.; Caballero-Marcos, A.; Sánchez-Martínez, A.; Di Maira, T.; Herrero, J.I.; Almohalla, C.; et al. Cumulative exposure to tacrolimus and incidence of cancer after liver transplantation. *Am. J. Transplant.* **2022**, *22*, 1671–1682. [CrossRef]
27. Durand, F. How to improve long-term outcome after liver transplantation? *Liver Int.* **2018**, *38*, 134–138. [CrossRef]
28. Rana, A.; Ackah, R.L.; Webb, G.J.; Halazun, K.J.; Vierling, J.M.; Liu, H.; Wu, M.-F.; Yoeli, D.; Kueht, M.; Mindikoglu, A.L.; et al. No Gains in Long-term Survival After Liver Transplantation Over the Past Three Decades. *Ann. Surg.* **2019**, *269*, 20–27. [CrossRef]
29. Rawla, P.; Sunkara, T.; Barsouk, A. Epidemiology of colorectal cancer: Incidence, mortality, survival, and risk factors. *Prz. Gastroenterol.* **2019**, *14*, 89–103. [CrossRef]
30. ZfKD. *Zentrum für Krebsregisterdaten (ZfKD) im Robert Koch-Institut (2019): Datensatz des ZfKD auf Basis der Epidemiologischen LandesKrebsRegisterDaten, Verfügbare Diagnosejahre bis 2016*; Version: Epi2018_1; ZfKD: Berlin, Germany, 2019. [CrossRef]
31. Altieri, M.; Sérée, O.; Lobbedez, T.; Segol, P.; Abergel, A.; Blaizot, X.; Boillot, O.; Boudjema, K.; Coilly, A.; Conti, F.; et al. Risk factors of de novo malignancies after liver transplantation: A French national study on 11004 adult patients. *Clin. Res. Hepatol. Gastroenterol.* **2021**, *45*, 101514. [CrossRef]
32. Moghimi-Dehkordi, B.; Safaee, A. An overview of colorectal cancer survival rates and prognosis in Asia. *World J. Gastrointest. Oncol.* **2012**, *4*, 71–75. [CrossRef]
33. Nitsche, U.; Maak, M.; Schuster, T.; Künzli, B.; Langer, R.; Slotta-Huspenina, J.; Janssen, K.-P.; Friess, H.; Rosenberg, R. Prediction of Prognosis Is Not Improved by the Seventh and Latest Edition of the TNM Classification for Colorectal Cancer in a Single-Center Collective. *Ann. Surg.* **2011**, *254*, 793–801. [CrossRef]
34. Washington, M.K. Colorectal carcinoma: Selected issues in pathologic examination and staging and determination of prognostic factors. *Arch. Pathol. Lab. Med.* **2008**, *132*, 1600–1607. [CrossRef] [PubMed]
35. Brenner, H.; Kloor, M.; Pox, C.P. Colorectal cancer. *Lancet* **2014**, *383*, 1490–1502. [CrossRef] [PubMed]
36. Gunderson, L.L.; Jessup, J.M.; Sargent, D.J.; Greene, F.L.; Stewart, A.K. Revised TN categorization for colon cancer based on national survival outcomes data. *J. Clin. Oncol.* **2010**, *28*, 264–271. [CrossRef] [PubMed]
37. Peruhova, M.; Peshevska-Sekulovska, M.; Panayotova, G.; Velikova, T. Foremost Concepts in Mechanisms of De Novo Post-Liver Transplantation Malignancy. *Gastroenterol. Insights* **2021**, *12*, 283–292. [CrossRef]
38. Guo, L.; Wang, C.; Qiu, X.; Pu, X.; Chang, P. Colorectal Cancer Immune Infiltrates: Significance in Patient Prognosis and Immunotherapeutic Efficacy. *Front. Immunol.* **2020**, *11*, 1052. [CrossRef]
39. Malka, D.; Lièvre, A.; André, T.; Taïeb, J.; Ducreux, M.; Bibeau, F. Immune scores in colorectal cancer: Where are we? *European J. Cancer* **2020**, *140*, 105–118. [CrossRef]
40. Luo, Z.; Chen, X.; Zhang, Y.; Huang, Z.; Zhao, H.; Zhao, J.; Li, Z.; Zhou, J.; Liu, J.; Cai, J.; et al. Development of a Metastasis-Related Immune Prognostic Model of Metastatic Colorectal Cancer and Its Usefulness to Immunotherapy. *Front. Cell Dev. Biol.* **2021**, *8*, 577125. [CrossRef]
41. Shaked, A.; DesMarais, M.R.; Kopetskie, H.; Feng, S.; Punch, J.D.; Levitsky, J.; Reyes, J.; Klintmalm, G.B.; Demetris, A.J.; Burrell, B.E.; et al. Outcomes of immunosuppression minimization and withdrawal early after liver transplantation. *Am. J. Transplant.* **2019**, *19*, 1397–1409. [CrossRef]
42. Girlanda, R.; Rela, M.; Williams, R.; O'Grady, J.G.; Heaton, N.D. Long-term outcome of immunosuppression withdrawal after liver transplantation. *Transplant Proc.* **2005**, *37*, 1708–1709. [CrossRef]
43. Duizendstra, A.A.; de Knegt, R.J.; Betjes, M.G.H.; Coenen, S.; Murad, S.D.; de Man, R.A.; Metselaar, H.J.; Sprengers, D.; Litjens, N.H.R.; Kwekkeboom, J. Immunosuppressive drug withdrawal late after liver transplantation improves the lipid profile and reduces infections. *Eur. J. Gastroenterol. Hepatol.* **2019**, *31*, 1444–1451. [CrossRef]
44. Toti, L.; Manzia, T.M.; Blasi, F.; Tisone, G. Withdrawal of Immunosuppression after Liver Transplantation. In *Liver Transplantation*; Wiley: Hoboken, NJ, USA, 2021; pp. 386–392.
45. Gonzalez, H.; Hagerling, C.; Werb, Z. Roles of the immune system in cancer: From tumor initiation to metastatic progression. *Genes Dev.* **2018**, *32*, 1267–1284. [CrossRef] [PubMed]
46. Bhatia, A.; Kumar, Y. Cancer Immunoediting: Immunosurveillance, Immune Equilibrium, and Immune Escape. In *Cancer Immunology: A Translational Medicine Context*; Rezaei, N., Ed.; Springer International Publishing: Cham, Switzerland, 2020; pp. 291–305.
47. Mohamed, I.B.; Aloor, F.Z.; Jalal, P.K. Strategies to Improve Immune Suppression Post-Liver Transplantation: A Review. *Transplantology* **2021**, *2*, 441–454. [CrossRef]
48. Londoño, M.-C.; Rimola, A.; O'Grady, J.; Sanchez-Fueyo, A. Immunosuppression minimization vs. complete drug withdrawal in liver transplantation. *J. Hepatol.* **2013**, *59*, 872–879. [CrossRef] [PubMed]
49. Cillo, U.; De Carlis, L.; Del Gaudio, M.; De Simone, P.; Fagiuoli, S.; Lupo, F.; Tisone, G.; Volpes, R. Immunosuppressive regimens for adult liver transplant recipients in real-life practice: Consensus recommendations from an Italian Working Group. *Hepatol. Int.* **2020**, *14*, 930–943. [CrossRef]
50. Nitta, H.; Younès, A.; El-Domiaty, N.; Karam, V.; Sobesky, R.; Vibert, E.; Coilly, A.; Maria Antonini, T.; De Martin, E.; Cherqui, D.; et al. High trough levels of everolimus combined to sorafenib improve patients survival after hepatocellular carcinoma recurrence in liver transplant recipients. *Transpl. Int.* **2021**, *34*, 1293–1305. [CrossRef] [PubMed]
51. Zou, Z.; Tao, T.; Li, H.; Zhu, X. mTOR signaling pathway and mTOR inhibitors in cancer: Progress and challenges. *Cell Biosci.* **2020**, *10*, 31. [CrossRef] [PubMed]

52. Wang, Y.J.; Chi, N.H.; Chou, N.K.; Huang, S.C.; Wang, C.H.; Wu, I.H.; Yu, H.Y.; Chen, Y.S.; Tsao, C.I.; Shun, C.T.; et al. Malignancy After Heart Transplantation Under Everolimus Versus Mycophenolate Mofetil Immunosuppression. *Transplant. Proc.* **2016**, *48*, 969–973. [CrossRef] [PubMed]
53. Kim, H.; Lee, K.W.; Oh, S.C.; Park, M.Y.; Seo, S.; Jin, X.L.; Hong, S.K.; Yoon, K.C.; Yi, N.J.; Suh, K.S. Sirolimus and MMF are insufficient immunosuppressants for regulation of the proliferation of CD133+EpCAM+ cell populations in HCC cell lines. *Biomed. Rep.* **2020**, *13*, 69. [CrossRef]
54. Fischer, L.; Saliba, F.; Kaiser, G.M.; De Carlis, L.; Metselaar, H.J.; De Simone, P.; Duvoux, C.; Nevens, F.; Fung, J.J.; Dong, G.; et al. Three-year Outcomes in De Novo Liver Transplant Patients Receiving Everolimus With Reduced Tacrolimus: Follow-Up Results From a Randomized, Multicenter Study. *Transplantation* **2015**, *99*, 1455–1462. [CrossRef]
55. Cholongitas, E.; Goulis, I.; Theocharidou, E.; Antoniadis, N.; Fouzas, I.; Giakoustidis, D.; Imvrios, G.; Giouleme, O.; Papanikolaou, V.; Akriviadis, E.; et al. Everolimus-based immunosuppression in liver transplant recipients: A single-centre experience. *Hepatol. Int.* **2014**, *8*, 137–145. [CrossRef]
56. Navarro-Villarán, E.; Tinoco, J.; Jiménez, G.; Pereira, S.; Wang, J.; Aliseda, S.; Rodríguez-Hernández, M.A.; González, R.; Marín-Gómez, L.M.; Gómez-Bravo, M.A.; et al. Differential Antitumoral Properties and Renal-Associated Tissue Damage Induced by Tacrolimus and Mammalian Target of Rapamycin Inhibitors in Hepatocarcinoma: In Vitro and In Vivo Studies. *PLoS ONE* **2016**, *11*, e0160979. [CrossRef] [PubMed]
57. Xie, Y.-H.; Chen, Y.-X.; Fang, J.-Y. Comprehensive review of targeted therapy for colorectal cancer. *Signal Transduct. Target. Ther.* **2020**, *5*, 22. [CrossRef] [PubMed]
58. Johdi, N.A.; Sukor, N.F. Colorectal Cancer Immunotherapy: Options and Strategies. *Front. Immunol.* **2020**, *11*, 1624. [CrossRef]
59. Sonbol, M.B.; Mountjoy, L.J.; Firwana, B.; Liu, A.J.; Almader-Douglas, D.; Mody, K.; Hubbard, J.; Borad, M.; Ahn, D.H.; Murad, M.H.; et al. The Role of Maintenance Strategies in Metastatic Colorectal Cancer: A Systematic Review and Network Meta-analysis of Randomized Clinical Trials. *JAMA Oncol.* **2020**, *6*, e194489. [CrossRef] [PubMed]
60. Adam, R.; Kitano, Y. Multidisciplinary approach of liver metastases from colorectal cancer. *Ann. Gastroenterol. Surg.* **2019**, *3*, 50–56. [CrossRef] [PubMed]
61. Lang, H. Liver resection is beneficial for patients with colorectal liver metastases and extrahepatic disease. *Ann. Transl. Med.* **2020**, *8*, 1122. [CrossRef]

medicina

MDPI

Article

Efficacy and Safety of Neoadjuvant Chemotherapy Combined with Adjuvant Chemotherapy for Locally Advanced Colon Cancer: A Propensity Score-Matching Analysis

Wei Zeng [1,†], Yi Liu [2,†], Chuandong Wang [1,†], Changshun Yang [3], Shengtao Lin [3] and Weihua Li [1,3,*]

1 Shengli Clinical Medical College of Fujian Medical University, Fuzhou 350001, China
2 Department of Endoscopy, National Cancer Center/National Clinical Research Center for Cancer/Cancer Hospital, Chinese Academy of Medical Science and Peking Union Medical College, Beijing 100021 China
3 Department of Surgical Oncology, Fujian Provincial Hospital, Fuzhou 350001, China
* Correspondence: liwh@fjmu.edu.cn
† These authors contributed equally to this work.

Abstract: *Background and Objectives*: Increasing evidence supports the use of neoadjuvant chemotherapy (NAC) for locally advanced colon cancer (LACC). However, its effectiveness remains controversial. This study explored the safety and efficacy of NAC combined with laparoscopic radical colorectal cancer surgery and adjuvant chemotherapy (AC) for LACC. *Materials and Methods*: We retrospectively analyzed 444 patients diagnosed with LACC (cT4 or cT3, with ≥5 mm invasion beyond the muscularis propria) in our hospital between 2012 and 2015. Propensity score matching (PSM; 1:2) was performed to compare patients treated with NAC and those treated with adjuvant chemotherapy (AC). *Results*: Overall, 42 patients treated with NAC were compared with 402 patients who received only AC. After PSM, 42 patients in the NAC group were compared with 84 patients in the control group, with no significant differences in the baseline characteristics between groups. The pathological tumor sizes in the NAC group were significantly smaller than those in the AC group (3.1 ± 2.1 cm vs. 5.8 ± 2.5 cm). Patients in the NAC group had a significantly lower T stage than those in the AC group ($p < 0.001$). After neoadjuvant chemotherapy, a significant response was observed in four (9.6%) patients, with two (4.8%) showing a complete response. The 5-year overall survival rates (88.1% vs. 77.8%, $p = 0.206$) and 5-year disease-free survival rates (75.1% vs. 64.2%, $p = 0.111$) did not differ between the groups. However, the 5-year cumulative rate of distant recurrence was significantly lower in the NAC than in the AC group (9.6% vs. 29.9%, $p = 0.022$). *Conclusions*: NAC, combined with AC, could downstage primary tumors of LACC and seems safe and acceptable for patients with LACC, with a similar long-term survival between the two treatments.

Keywords: neoadjuvant chemotherapy (NAC); locally advanced colon cancer (LACC); preoperative treatment; propensity score matching; survival

Citation: Zeng, W.; Liu, Y.; Wang, C.; Yang, C.; Lin, S.; Li, W. Efficacy and Safety of Neoadjuvant Chemotherapy Combined with Adjuvant Chemotherapy for Locally Advanced Colon Cancer: A Propensity Score-Matching Analysis. *Medicina* **2022**, *58*, 1505. https://doi.org/10.3390/medicina58111505

Academic Editors: Antonio M Scanu, Maria Rosaria De Miglio and Seung-Gu Yeo

Received: 17 September 2022
Accepted: 17 October 2022
Published: 22 October 2022

1. Introduction

Colon cancer (CC) is the fourth most common type of cancer worldwide [1]. Among patients with CC, a substantial proportion that presents with locally advanced colon cancer (LACC) (T4 or T3, with ≥5 mm invasion beyond the muscularis propria) still have an unsatisfactory prognosis, with 5-year survival rates ranging from 55% to 88%, despite developments in surgical technique and chemotherapy regimens [2]. Worldwide, the current standard treatment strategy for LACC is radical surgical resection of the tumor (R0 resection), followed by adjuvant chemotherapy. Regarding the clinical treatment strategy of other solid tumors, neoadjuvant chemotherapy (NAC) has been successfully applied in the clinical treatment of cancers, including rectal and breast cancer [3–5]. Relevant research has suggested that neoadjuvant chemotherapy (NAC) was useful in promoting a reduction in tumor burden prior to surgery and the eradication of micro-metastases [6], which achieved a higher rate of R0 resection. However, it usually takes about one month for patients to fully recover from surgery and receive adjuvant chemotherapy (AC). A previous study recognized that metabolic activity increased after surgical removal of the primary tumor, suggesting that surgical stimulation of growth factors may be one of the factors promoting postoperative metastasis [7]. From this perspective, preoperative NAC may have a positive impact on patient prognosis [8]. Therefore, new treatment strategies urgently need to be proposed and validated.

However, there are still relatively few studies showing the usefulness of NAC for survival in patients with LACC. Most recent studies have focused on demonstrating the feasibility and safety of NAC [9–11]. The FOxTROT study showed that preoperative chemotherapy combining 5-fluorouracil and oxaliplatin with or without panitumumab in patients with resectable T4 or T3 colon cancer had a significant effect on tumor downstaging and had high safety [12]. The ongoing French clinical trial PRODIGE 22-ECKINOXE and the Chinese COLARC study have both confirmed that NAC is feasible, with acceptable tolerability, but is not associated with an increased major pathological response rate [13–15]. A retrospective study also showed that patients with clinical T4b CC treated with NAC might have an improved survival rate [16], but this has not been observed in patients with clinical T3 or T4a CC. In terms of clinical guidelines, the current National Comprehensive Cancer Network (NCCN) guidelines also recommend preoperative neoadjuvant chemoradiotherapy as an option for patients with initially unresectable, non-metastatic T4 colon cancer [17].

The application of NAC for LACC is challenged by concerns that patients may lose the opportunity to undergo radical surgery due to the progression of the primary tumor during NAC, while some patients may receive over-treatment due to inaccurate computed tomography (CT) staging. With the advancements in CT, many studies have confirmed the accuracy of CT technology in staging CC [18–20], and CT scanning can accurately identify high-risk (T3/4) colon cancers with minimal over-staging of T1/T2 tumors [21]. Given the potential advantages and disadvantages, we conducted this study to investigate the perioperative efficacy and postoperative outcomes to evaluate whether NAC could improve prognosis in patients with LACC and who only received surgery combined with postoperative AC for the time being.

2. Materials and Methods

2.1. Patient Selection

We reviewed the data of 444 patients with LACC (T4 or T3, with ≥5 mm invasion beyond the muscularis propria) who underwent surgery at the Fujian Provincial Hospital between 2012 and 2015 (Figure 1). Patients were randomly assigned to two groups: 42 received preoperative NAC combined with AC, while the remaining 402 received postoperative AC. The inclusion criteria were as follows: (1) histologically confirmed colon cancer; (2) CT-verified colon cancer at clinical stage T4a or T3, with ≥5 mm invasion beyond the muscularis propria; and (3) radical surgery. The exclusion criteria were as follows: (1) distant metastasis detected upon preoperative examination; (2) simultaneous malignancies from other organs or prior malignancy; (3) serious cardiovascular or cerebrovascular diseases, liver and kidney dysfunction, severe blood system diseases, immune system diseases, or severe mental disorder; (4) incomplete or inaccurate medical records; and (5) below 18 years old or over 90 years old at the time of diagnosis. This study was reviewed and approved by the Ethics Committee of the Fujian Provincial Hospital and was registered under the ethics committee approval number K2017-09-070. All data were anonymized, and the requirement for informed consent was therefore waived. All study procedures were performed in accordance with the Helsinki Declaration of 1964 and its later versions.

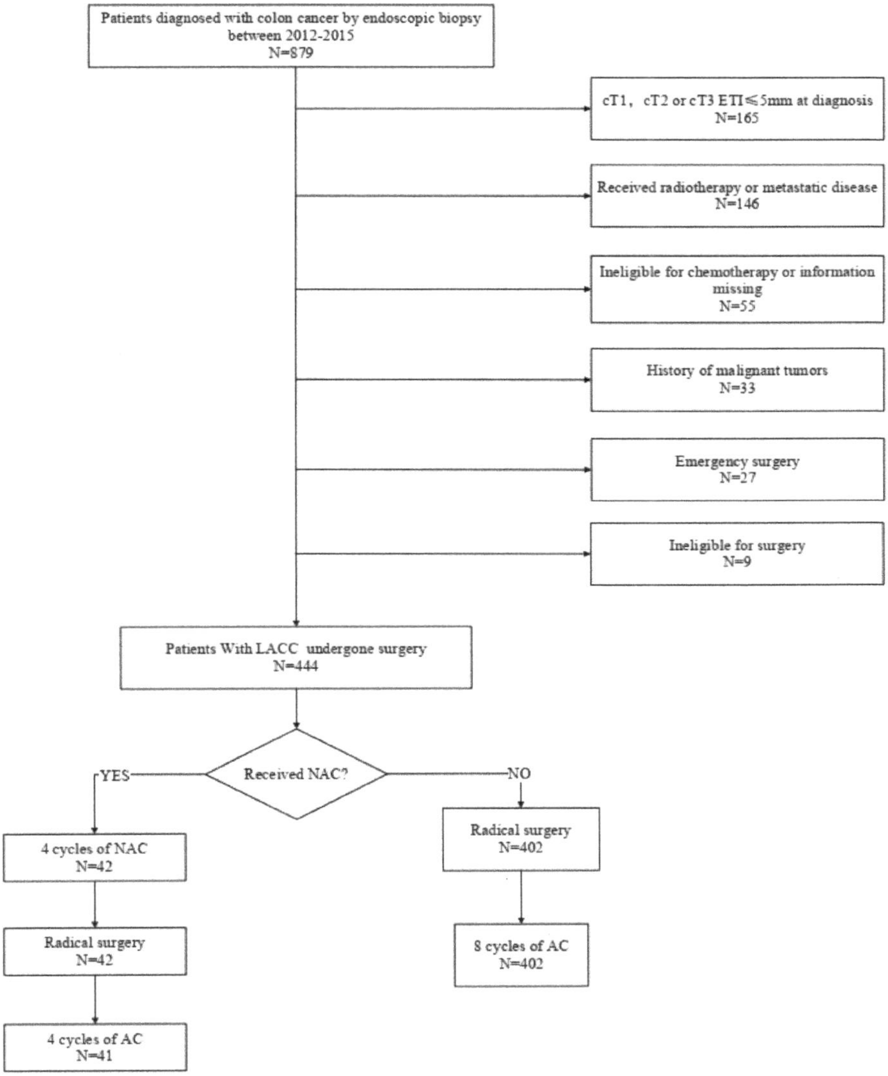

Figure 1. Flowchart of patient selection.

2.2. Treatment Regimes

Initial clinical staging using colonoscopy with biopsy confirmation and abdominal computed tomography (CT) was performed in all cases. Patients in the NAC group received 6 cycles of XELOX (capecitabine 1000 mg/m^2 orally days 1–14 q3w, oxaliplatin 130 mg/m^2 iv day 1 q3w) after diagnosis and underwent radical surgery three weeks after the last cycle of NAC. The response to NAC was assessed every three cycles by performing a CT scan (according to RECIST [22]) and measuring serum carcinoembryonic antigen (CEA) and carbohydrate antigen199 (CA199) levels. Further AC was determined based on the pathological results and the patient's willingness to undergo the remaining two cycles of XELOX. For patients in the AC group, radical surgery was performed first after diagnosis, and patients received eight cycles of AC (XELOX, capecitabine 1000 mg/m^2 orally days

1–14 q3w, oxaliplatin 130 mg/m^2 iv day 1 q3w), depending on the histological stage and the pathological response. Follow-up was performed 1 month after surgery, every 3 months for 3 years, every 6 months for 5 years, and yearly thereafter.

2.3. Data Collection

The following variables were included in the analysis: sex, age, tumor site, tumor size, gross type, tumor differentiation, histopathology, clinical T and N stages, serum CEA and CEA levels, ASA grade, body mass index (BMI), operation time, estimated blood loss, time to start the diet, length of hospital stay, toxic effect, pathologic outcomes according to the American Joint Committee on Cancer (AJCC) guidelines (8th edition), and postoperative morbidity and mortality. Toxicity was assessed according to the Common Toxicity Criteria for Adverse Events (version 3.0). The date of diagnosis was defined as the date of the first histological confirmation of malignancy, most often the day of the endoscopic biopsy. After resection, the pathologist performed the final stage. A pathological tumor (ypT) and nodal staging were compared with clinical staging in both groups to assess the downstaging effects of neoadjuvant CT. R0 resection was achieved if the resection margins were microscopically tumor-free. In the case of irradical resection, the resection was either labeled R1 (microscopic involvement of the resection margins) or R2 (macroscopic involvement). Major postoperative complications such as wound infection, ileus, and anastomotic leakage were recorded. The primary outcome was overall survival. The secondary endpoints were recurrence rate, disease-free survival (DFS), and chemotherapy toxicity.

2.4. Statistical Analysis

The chi-square test or Fisher's exact probability method was used to compare classified variables between the two groups. An independent-samples *t*-test was used to compare normally distributed continuous variables. Nonparametric Mann–Whitney U tests were applied when the variance was not normally distributed. Propensity score matching was applied to reduce the possibility of selection bias and adjust for significant differences in the baseline characteristics of the patients. The propensity score was calculated based on sex, age, tumor site, tumor size, gross type, tumor differentiation, histopathology, clinical T and N stage, CEA levels, American Society of Anesthesiologists (ASA) grade, and body mass index (BMI). Patients in the NAC group were matched 1:2 using nearest neighbor matching based on the closest propensity score to those in the AC group. Overall survival and disease-free survival rates were estimated using the Kaplan–Meier method. Statistical significance was set at $p < 0.05$. Statistical analysis was performed using SPSS for Windows (version 22.0; SPSS Inc., Chicago, IL, USA).

3. Results

3.1. Baseline Characteristics

This study included 444 patients with LACC who underwent radical surgical resection between January 2012 and June 2015. Among them, 42 patients received NAC before surgery, while the remaining 402 patients underwent surgical resection without preoperative chemotherapy. Before propensity score matching, sex, gross type, tumor differentiation, histopathology, cT and cN stages, serum CEA level, serum CA199 level, ASA, and BMI were not significantly different between the groups (Table 1). However, compared to the AC group, patients in the NAC group were significantly older (66.48 ± 11.98 years vs. 61.60 ± 13.68 years, $p = 0.027$), the tumor sizes were significantly larger (5.0 ± 1.7 vs. 4.2 ± 2.0 cm, $p = 0.009$), and more tumors were located in the left colon ($p = 0.015$). A propensity score was calculated to adjust for biases caused by differences in baseline characteristics between the two groups. After matching, there were no significant differences in any baseline characteristics between the groups (Table 1). We found that several indices of patients in the NAC group, such as body mass index (BMI), serum CEA level, serum CA199 level, and American Society of Anesthesiologists (ASA) score, improved after six cycles of NAC. (Table 2, $p < 0.05$).

Table 1. Overall patient and tumor characteristics before and after PSM for the NAC and AC groups.

Variable	Raw Data		p	After Propensity Matching		p
	NAC (n = 42)	AC (n = 402)		NAC (n = 42)	AC (n = 84)	
Gender			0.612			0.377
Male	22	227		22	37	
Female	20	175		20	47	
Age, years			0.027			0.670
Mean ± SD (range)	66.48 ± 11.98	61.60 ± 13.68		66.48 ± 11.98	65.51 ± 11.95	
Tumor site			0.015			0.777
Right colon	12	194		12	22	
Left colon	30	208		30	62	
Tumor size, cm (imaging)			0.009			0.635
Mean ± SD	5.0 ± 1.7	4.2 ± 2.0		5.0 ± 1.7	4.8 ± 2.0	
Morphology			0.855			0.077
Infiltrative	1	23		1	0	
Ulcerative	27	247		27	43	
Expanding	14	132		14	41	
Tumor differentiation			0.887			0.280
Well or moderately	34	329		34	74	
Poorly, others	8	73		8	10	
Histopathology			0.847			0.541
Tubular adenocarcinoma	34	328		34	74	
Mucinous adenocarcinoma	6	55		6	7	
Signet ring cell carcinoma	0	6		0	1	
Others	2	13		2	2	
cT stage *			0.436			0.172
cT3	19	157		19	27	
cT4	23	245		23	57	
cN stage *			0.214			0.591
cN0	20	213		20	38	
cN1	10	111		10	28	
cN2	9	69		9	15	
cNx	3	9		3	3	
CEA, ng/ml			0.093			0.074
Median (P_{25}, P_{75})	5.62 (2.81, 13.63)	4.94 (2.08, 13.10)		5.62 (2.81, 13.63)	4.82 (2.00, 12.68)	
CA199, U/mL			0.328			0.362
Median (P_{25}, P_{75})	19.63 (11.97, 27.75)	17.02 (8.38, 27.84)		19.63 (11.97, 27.75)	16.10 (10.03, 27.40)	

Table 1. *Cont.*

Variable	Raw Data		p	After Propensity Matching		p
	NAC (n = 42)	AC (n = 402)		NAC (n = 42)	AC (n = 84)	
ASA			0.333			0.960
I	26	203		26	53	
II	13	126		13	24	
III	3	67		3	7	
IV	0	6		0	0	
BMI			0.269			0.153
Mean ± SD (range)	20.64 ± 4.37	21.51 ± 4.88		20.64 ± 4.37	21.85 ± 4.49	

* According to the AJCC Cancer Staging Manual, 8th edition. Abbreviations: ASA, American Society of Anesthesiologists; BMI, body mass index.

Table 2. Patient characteristics after neoadjuvant chemotherapy for the NAC group.

Variable	NAC-Before (n = 42)	NAC-After (n = 42)	p
CEA, ng/mL			0.003
Median (P_{25}, P_{75})	5.62 (2.81, 13.63)	3.25 (2.66, 4.20)	
CA199, U/ml			0.001
Median (P_{25}, P_{75})	19.63 (11.97, 27.75)	12.42 (4.55, 1.40)	
ASA			0.009
I	26	37	
II	13	5	
III	3	0	
IV	0	0	
BMI			0.001
Mean ± SD	20.65 ± 4.37	23.46 ± 3.28	

Abbreviations: NAC-after, patients of the NAC group after 6 cycles of neoadjuvant chemotherapy; NAC-before, patients of the NAC group before neoadjuvant chemotherapy; ASA, American Society of Anesthesiologists; BMI, body mass index.

3.2. Perioperative Outcomes

The operation time, estimated blood loss, time to bowel movement, time to a liquid diet, time to a soft diet, postoperative hospital stays, and complications within 30 days of surgery were similar between the two groups (Table 3). In addition, there was no significant difference in mortality between the two groups 30 days after surgery. Regarding the toxic effects of chemotherapy, there was no significant difference in the incidence of gastrointestinal, hematologic, and dermatologic effects; however, the NAC group had a lower incidence of any grade 3 or 4 toxic effects than the AC group (10.0% vs. 25.9%, p = 0.041). Four (9.5%) and eighteen (21.4%) patients did not complete the full cycles of chemotherapy in the NAC and AC groups due to toxic effects, respectively.

Table 3. Comparison of perioperative outcomes between NAC and AC groups.

Variable	NAC (n = 42)	AC (n = 84)	p
Operation time, min			0.183
Median (P_{25}, P_{75})	185.50 (165.50, 201.00)	175.50 (147.25, 200.75)	
Estimated blood loss, ml			0.111
Median (P_{25}, P_{75})	50.00 (35.00, 60.00)	55.00 (40.00, 65.00)	
Anal exhaust time, day			0.757
Median (P_{25}, P_{75})	3.00 (2.00, 3.00)	3.00 (2.00, 3.00)	
Time to liquid diet, day			0.375
Median (P_{25}, P_{75})	2.00 (1.00, 2.00)	1.50 (1.00, 2.00)	
Time to soft diet, day			0.383
Median (P_{25}, P_{75})	3.00 (3.00, 4.00)	4.00 (3.00, 4.00)	
Postoperative hospital stays, day			0.419
Median (P_{25}, P_{75})	6.00 (5.00, 8.00)	6.00 (5.00, 7.00)	
Complication within 30 days of surgery			1.000
None	33	67	
Wound infection	2	5	
Ileus	5	9	
Anastomotic leakage	2	3	
Mortality within 30 days of surgery			0.552
No	42	82	
Yes	0	2	
Toxic effect *			
Gastrointestinal **	7	19	0.436
Hematologic effects	10	32	0.109
Dermatologic effects	9	25	0.321
Any grade 3 or 4 toxic effect	4	22	0.041

* According to National Cancer Institute Common Toxicity Criteria; ** Nausea, vomiting, and diarrhea.

3.3. Pathological Outcomes

None of the patients experienced progression during neoadjuvant chemotherapy, and the NAC group achieved a smaller tumor size than the AC group (3.1 ± 2.1 vs. 5.8 ± 2.5 cm, $p < 0.001$). In all patients, the cT stage was reported before the start of NAC. Four patients showed significant downstaging of the primary tumor after systemic therapy (cT3-4 to pT0-2, 9.5%), while two patients showed a complete pathological response (pT0; Table 4). None of the patients in the NAC group had nodal over-staging. Although only three patients (21.4%) were diagnosed with cN1and finally had pN2 disease, up to 15 patients (65.2%) were diagnosed with cN0 and finally had pN1-2 disease (Table 5).

Table 4. Comparison of pathologic outcomes between the NAC and AC groups.

Variable	NAC (n = 42)	AC (n = 84)	p
Tumor size, cm (pathological)			<0.001
Mean ± SD	3.1 ± 2.1	5.8 ± 2.5	
T stage *			<0.001
T0	2	0	
T1	2	0	
T2	8	0	
T3	23	29	
T4	7	55	
N stage *			0.310
N0	22	49	
N1	10	24	
N2	10	11	
Resection margin			1.000
R0	40	81	
R1	2	3	
Angiolymphatic invasion			0.725
Positive	35	72	
Negative	7	12	
Nerve invasion			1.000
Positive	40	81	
Negative	2	3	

* According to the AJCC Cancer Staging Manual, 8th edition.

Table 5. Clinical and pathological nodal staging.

(a) Nodal downstaging in patients who received NAC				
		(a)		
	Pathological N-score			
	pN0	pN1	pN2	Total
Clinical N-score				42
cN0	20	0	0	20
cN1	0	10	0	10
cN2	0	0	9	9
cNx	2	0	1	3
Total	22	10	10	42

Table 5. *Cont.*

(b) Comparison of clinical and pathological nodal staging in patients treated with AC.				
	(b)			
	Pathological N-score			
	pN0	**pN1**	**pN2**	**Total**
Clinical N-score				
cN0	23	9	6	38
cN1	14	11	3	28
cN2	10	3	2	15
cNx	2	1	0	3
Total	49	24	11	84

Abbreviations: NAC, neoadjuvant chemotherapy; AC, adjuvant chemotherapy.

3.4. Survival

The median follow-up periods in the NAC and AC groups were 56 (12–80) and 66.5 (2–83) months, respectively, while the corresponding 5-year overall survival rates were 88.1% and 77.8%, respectively. This difference was not significant (p = 0.206; Figure 2a). Furthermore, there was no significant difference in the 5-year progression-free survival between the two groups (75.1% vs. 64.2%, p = 0.111; Figure 2b), nor in the incidence of 5-year local recurrence (18.3% vs. 15.3%; p = 0.935; Figure 2c). However, the 5-year cumulative incidence of distal recurrence was 9.6% in the NAC group, which was significantly lower than that in the AC group (29.9%, p = 0.018; Figure 2d).

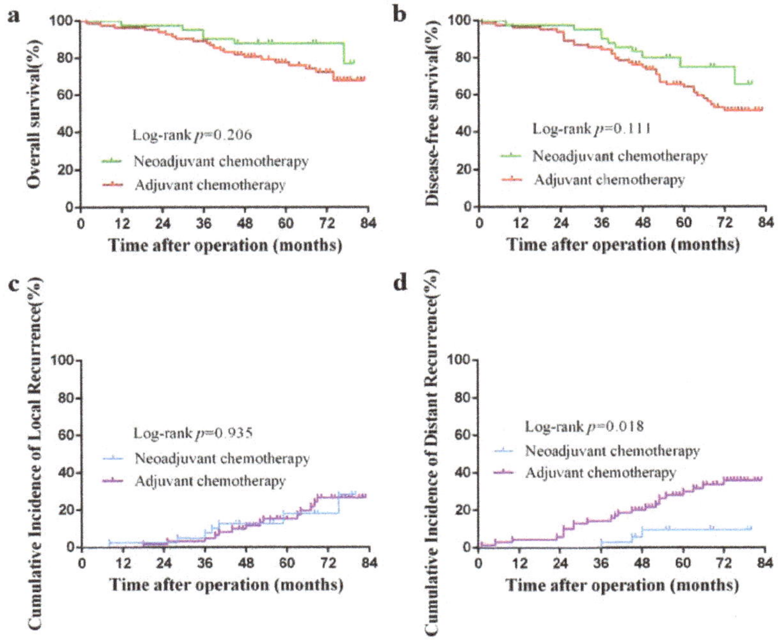

Figure 2. (**a**) Kaplan–Meier curve for overall survival after propensity score matching; (**b**) Kaplan–Meier curve for disease-free survival after propensity score matching; (**c**) Kaplan–Meier curve for cumulative incidence of local recurrences; (**d**) Kaplan–Meier curve for cumulative incidence of distant recurrences.

4. Discussion

NAC for LACC has become increasingly frequently applied in the clinic; however, its applicability remains controversial [6,9,11,16]. Our study illustrated that NAC combined with AC was not only safe but also resulted in significant tumor downstaging in patients with LACC, and the long-term outcomes were similar to those of patients who underwent surgery directly after diagnosis.

It is well known that chemotherapy drugs induce certain toxicity towards the liver and kidney, and it has been suggested that NAC may be associated with unnecessary patient morbidity due to chemotherapeutic toxicities. When comparing patients who received NAC with those who did not, we found that patients in the NAC group had a lower incidence of grade 3 or 4 toxicities. Clinically, neoadjuvant therapy toxicity (grade 3 or 4) was observed in only 10% of the patients in the NAC group. Interestingly, our further research suggested that several indices of patients in the NAC group, such as serum CEA level, serum CA199 level, BMI, and ASA score, were improved after six cycles of NAC. This may be because NAC was usually carried out before radical surgery, delaying the operation time by about 12–18 weeks. In addition, we increased the nutritional intake through enteral and parenteral nutrition during chemotherapy. As a result, patients had the opportunity to improve their physical condition before surgery. Patients tolerated b preoperative chemotherapy better than postoperative adjuvant chemotherapy because they were in a relatively healthier state.

A previous study indicated that the rates of adverse reactions and surgical complications did not differ between patients who underwent NAC and those who did not. Karoui et al. and the FOxTROT Collaborative Group both demonstrated that there was no significant difference in postoperative anastomotic leaks, wound infections, or return to the theater between the neoadjuvant and control arms in both RCTs [12,15]. The results of our study suggest that operation time, estimated blood loss, time to bowel movement, time to a liquid diet, time to a soft diet, postoperative hospital stays, and mortality within 30 days of surgery did not show any statistical difference between the NAC and AC groups. In addition, the occurrence of major complications, such as wound infection, ileus, and anastomotic leakage, was equal between the groups. The results of this study clearly showed that NAC is well tolerated with an acceptable side effect profile for an average of less than 30 days after surgery. Thus, we supposed that NAC was non-inferior in terms of safety and did not increase surgical complications or mortality compared to standard surgery. Other concerns raised about NAC were related to the possibility that the response of tumors to neoadjuvant therapies remains variable; a subgroup of patients may not achieve any downstaging of the tumor, and some of them may even show disease progression, as observed in locally advanced rectal cancer due to delays in operative intervention [23,24]. However, it was encouraging that no progression was observed during neoadjuvant chemotherapy in this study. This finding is in agreement with the results of a previous study [12].

The subjects included in our present study were patients with LACC, such as T4 or high-risk T3 (with $\geq$5 mm invasion beyond the muscularis propria), without distant metastases. Identifying this patient population relied heavily on accurate CT staging, as it guided the need for neoadjuvant therapy. CT staging was found to be accurate, with an overall sensitivity of 90% in detecting tumor invasion beyond the bowel wall and nodal involvement in a previous meta-analysis [25]. In this study, we enrolled 23 patients diagnosed with the cT4 stage in the NAC group. Tumor grade regression of the specimen is an important factor directly related to chemotherapy response [26,27]. The results of our study showed that the benefits of NAC included the significant downsizing of the primary tumor and downstaging of the T stage. However, several studies have indicated that the complete pathological response rate of LACC was between 2–4.6%, which is significantly lower than that of rectal cancers, which ranged from 15% to 25% [10,28,29]. In our study, the sizes of primary tumors were markedly reduced after NAC in 42 patients. Additionally, evidence of significant downstaging (cT3-4 to pT0-2, 9.5%) was demonstrated in 9.5% of patients, and a complete pathological response (pT0) was observed in 4.8% of patients. This is also in agreement with the results of a previous study [11].

Recently, a systematic review and meta-analysis concluded that NAC could significantly improve disease-free survival and overall survival in patients with rectal cancer [30]. Similarly, Cheong et al. found that patients with colon cancer receiving NAC also had better overall survival and disease-free survival [31]. In contrast, several studies have suggested that the overall survival of patients receiving NAC was similar to that of patients without NAC [11,16]. Our research showed no significant difference in overall survival and disease-free survival between the NAC and AC groups; however, NAC significantly reduced the incidence of distant recurrence. This may be explained by the fact that circulating tumor cells and lymph node metastasis could be eradicated by early systemic NAC. Furthermore, NAC may shrink tumors and reduce tumor cell shedding caused by surgical trauma. A related study showed that surgery stimulates growth factors and induces immunosuppression, which may promote tumor progression and the spread of micrometastases in the postoperative setting [13]. Surgery after NAC can remove the tumor more radically and eradicate systemic micrometastases earlier. This may prevent the occurrence of distant relapses.

Despite these positive findings, this study had several limitations. First, the sample size was relatively small. Second, selection bias could have occurred in the control group because only patients who were able to undergo adjuvant CT were included, and patients who died postoperatively or had severe complications were excluded. Third, although propensity score matching was performed to balance the significant baseline characteristics of patients, RCTs nevertheless need to be conducted to confirm our results.

5. Conclusions

Overall, our findings showed that a lower incidence of any grade 3 or 4 toxic effects were observed in the NAC group, and there was no significant increase in postoperative complications or mortality. NAC combined with AC could be used to downstage the primary tumor of the LACC and eliminate potential micrometastases. NAC combined with AC appears to be a safe and acceptable modality for patients with LACC. However, additional large randomized trials with longer follow-up times are needed to provide more reliable results.

Author Contributions: Conceptualization, W.L. and W.Z.; methodology, W.Z.; software, W.Z.; validation, W.Z., Y.L. and C.W.; formal analysis, W.Z.; investigation, W.Z.; resources, Y.L.; data curation, C.W.; writing—original draft preparation, W.Z., Y.L. and C.W.; writing—review and editing, W.Z., Y.L. and C.W.; visualization, W.Z., Y.L. and C.W.; supervision, C.Y. and S.L.; project administration, C.Y. and S.L; funding acquisition, W.L. All authors have read and agreed to the published version of the manuscript.

Funding: This research received no external funding.

Institutional Review Board Statement: The study was conducted in accordance with the Declaration of Helsinki, and approved by the Ethics Committee of the Fujian Provincial Hospital (protocol code K2017-09-070).

Informed Consent Statement: Informed consent was obtained from all subjects involved in the study.

Data Availability Statement: Please contact the corresponding author (Weihua Li, email: liwh@fjmu.edu.cn) for data requests.

Acknowledgments: The authors thank all colleagues and nurses of the Department of Surgical Oncology who provided care to the patients in this study and the participants of the Department of Pathology for their contributions to this project.

Conflicts of Interest: The authors declare no conflict of interest.

References

1. Bray, F.; Ferlay, J.; Soerjomataram, I.; Siegel, R.L.; Torre, L.A.; Jemal, A. Global cancer statistics 2018: GLOBOCAN estimates of incidence and mortality worldwide for 36 cancers in 185 countries. *CA A Cancer J. Clin.* **2018**, *68*, 394–424. [CrossRef] [PubMed]
2. Lombardi, L.; Morelli, F.; Cinieri, S.; Santini, D.; Silvestris, N.; Fazio, N.; Orlando, L.; Tonini, G.; Colucci, G.; Maiello, E. Adjuvant colon cancer chemotherapy: Where we are and where we'll go. *Cancer Treat. Rev.* **2010**, *36*, S34–S41. [CrossRef]
3. Cunningham, D.; Allum, W.H.; Stenning, S.P.; Thompson, J.N.; Van de Velde, C.J.; Nicolson, M.; Scarffe, J.H.; Lofts, F.J.; Falk, S.J.; Iveson, T.J. Perioperative chemotherapy versus surgery alone for resectable gastroesophageal cancer. *N. Engl. J. Med.* **2006**, *355*, 11–20. [CrossRef]
4. Sauer, R.; Becker, H.; Hohenberger, W.; Rodel, C.; Wittekind, C.; Fietkau, R.; Martus, P.; Tschmelitsch, J.; Hager, E.; Hess, C.F.; et al. Preoperative versus postoperative chemoradiotherapy for rectal cancer. *N. Engl. J. Med.* **2004**, *351*, 1731–1740. [CrossRef]
5. Graham, P.J.; Brar, M.S.; Foster, T.; McCall, M.; Bouchard-Fortier, A.; Temple, W.; Quan, M.L. Neoadjuvant Chemotherapy for Breast Cancer, Is Practice Changing? A Population-Based Review of Current Surgical Trends. *Ann. Surg. Oncol.* **2015**, *22*, 3376–3382. [CrossRef] [PubMed]
6. Arredondo, J.; Pastor, C.; Baixauli, J.; Rodríguez, J.; González, I.; Vigil, C.; Chopitea, A.; Hernández-Lizoáin, J.L. Preliminary outcome of a treatment strategy based on perioperative chemotherapy and surgery in patients with locally advanced colon cancer. *Color. Dis.* **2013**, *15*, 552–557. [CrossRef]
7. Scheer, M.G.; Stollman, T.H.; Vogel, W.V.; Boerman, O.C.; Oyen, W.J.; Ruers, T.J. Increased metabolic activity of indolent liver metastases after resection of a primary colorectal tumor. *J. Nucl. Med.* **2008**, *49*, 887–891. [CrossRef]
8. van der Bij, G.J.; Oosterling, S.J.; Beelen, R.H.; Meijer, S.; Coffey, J.C.; van Egmond, M. The perioperative period is an underutilized window of therapeutic opportunity in patients with colorectal cancer. *Ann. Surg.* **2009**, *249*, 727–734. [CrossRef] [PubMed]
9. Zhou, Z.; Nimeiri, H.S.; Benson, A.B., 3rd. Preoperative chemotherapy for locally advanced resectable colon cancer—A new treatment paradigm in colon cancer? *Ann. Transl. Med.* **2013**, *1*, 11. [PubMed]
10. Jakobsen, A.; Andersen, F.; Fischer, A.; Jensen, L.H.; Jørgensen, J.C.R.; Larsen, O.; Lindebjerg, J.; Pløen, J.; Rafaelsen, S.; Vilandt, J. Neoadjuvant chemotherapy in locally advanced colon cancer. *A phase II Trial. Acta Oncol.* **2015**, *54*, 1747–1753. [PubMed]
11. de Gooyer, J.M.; Verstegen, M.G.; Radema, S.A.; Verhoeven, R.H.; Verhoef, C.; Schreinemakers, J.M.; de Wilt, J.H. Neoadjuvant Chemotherapy for Locally Advanced T4 Colon Cancer: A Nationwide Propensity-Score Matched Cohort Analysis. *Dig. Surg.* **2020**, *37*, 292–301. [CrossRef]
12. Group, F.O.C. Feasibility of preoperative chemotherapy for locally advanced, operable colon cancer: The pilot phase of a randomised controlled trial. *Lancet Oncol.* **2012**, *13*, 1152–1160.
13. Karoui, M.; Rullier, A.; Luciani, A.; Bonnetain, F.; Auriault, M.-L.; Sarran, A.; Monges, G.; Trillaud, H.; Le Malicot, K.; Leroy, K.; et al. Neoadjuvant FOLFOX 4 versus FOLFOX 4 with Cetuximab versus immediate surgery for high-risk stage II and III colon cancers: A multicentre randomised controlled phase II trial–the PRODIGE 22–ECKINOXE trial. *BMC Cancer* **2015**, *15*, 511. [CrossRef] [PubMed]
14. Liu, F.; Yang, L.; Wu, Y.; Li, C.; Zhao, J.; Keranmu, A.; Zheng, H.; Huang, D.; Wang, L.; Tong, T.; et al. CapOX as neoadjuvant chemotherapy for locally advanced operable colon cancer patients: A prospective single-arm phase II trial. *Chin. J. Cancer Res.* **2016**, *28*, 589–597. [CrossRef]
15. Karoui, M.; Rullier, A.; Piessen, G.; Legoux, J.L.; Barbier, E.; De Chaisemartin, C.; Lecaille, C.; Bouche, O.; Ammarguellat, H.; Brunetti, F.; et al. Perioperative FOLFOX 4 Versus FOLFOX 4 Plus Cetuximab Versus Immediate Surgery for High-Risk Stage II and III Colon Cancers: A Phase II Multicenter Randomized Controlled Trial (PRODIGE 22). *Ann. Surg.* **2020**, *271*, 637–645. [CrossRef] [PubMed]
16. Dehal, A.; Graff-Baker, A.N.; Vuong, B.; Fischer, T.; Klempner, S.J.; Chang, S.C.; Grunkemeier, G.L.; Bilchik, A.J.; Goldfarb, M. Neoadjuvant Chemotherapy Improves Survival in Patients with Clinical T4b Colon Cancer. *J. Gastrointest. Surg.* **2018**, *22*, 242–249. [CrossRef] [PubMed]
17. Azab, B.; Bhatt, V.; Phookan, J.; Murukutla, S.; Kohn, N.; Terjanian, T.; Widmann, W.D. Usefulness of the neutrophil-to-lymphocyte ratio in predicting short- and long-term mortality in breast cancer patients. *Ann. Surg. Oncol.* **2012**, *19*, 217–224. [CrossRef]
18. Dighe, S.; Blake, H.; Koh, M.; Swift, I.; Arnaout, A.; Temple, L.; Barbachano, Y.; Brown, G. Accuracy of multidetector computed tomography in identifying poor prognostic factors in colonic cancer. *Br. J. Surg.* **2010**, *97*, 1407–1415. [CrossRef] [PubMed]
19. Dighe, S.; Purkayastha, S.; Swift, I.; Tekkis, P.; Darzi, A.; A'Hern, R.; Brown, G. Diagnostic precision of CT in local staging of colon cancers: A meta-analysis. *Clin. Radiol.* **2010**, *65*, 708–719. [CrossRef]
20. Smith, N.J.; Bees, N.; Barbachano, Y.; Norman, A.R.; Swift, R.I.; Brown, G. Preoperative computed tomography staging of nonmetastatic colon cancer predicts outcome: Implications for clinical trials. *Br. J. Cancer* **2007**, *96*, 1030–1036. [CrossRef]
21. Dighe, S.; Swift, I.; Magill, L.; Handley, K.; Gray, R.; Quirke, P.; Morton, D.; Seymour, M.; Warren, B.; Brown, G. Accuracy of radiological staging in identifying high-risk colon cancer patients suitable for neoadjuvant chemotherapy: A multicentre experience. *Color. Dis.* **2012**, *14*, 438–444. [CrossRef]
22. Eisenhauer, E.A.; Therasse, P.; Bogaerts, J.; Schwartz, L.H.; Sargent, D.; Ford, R.; Dancey, J.; Arbuck, S.; Gwyther, S.; Mooney, M.; et al. New response evaluation criteria in solid tumours: Revised RECIST guideline (version 1.1). *Eur. J. Cancer* **2009**, *45*, 228–247. [CrossRef] [PubMed]

23. Travaini, L.L.; Zampino, M.G.; Colandrea, M.; E Ferrari, M.; Gilardi, L.; Leonardi, M.C.; Santoro, L.; Orecchia, R.; Grana, C.M. PET/CT with Fluorodeoxyglucose During Neoadjuvant Chemoradiotherapy in Locally Advanced Rectal Cancer. *Ecancermedicalscience* **2016**, *10*, 629. [CrossRef]
24. Gollins, S.; Sebag-Montefiore, D. Neoadjuvant Treatment Strategies for Locally Advanced Rectal Cancer. *Clin. Oncol.* **2016**, *28*, 146–151. [CrossRef] [PubMed]
25. Nerad, E.; Lahaye, M.J.; Maas, M.; Nelemans, P.; Bakers, F.C.H.; Beets, G.L.; Beets-Tan, R.G.H. Diagnostic Accuracy of CT for Local Staging of Colon Cancer: A Systematic Review and Meta-Analysis. *Am. J. Roentgenol.* **2016**, *207*, 984–995. [CrossRef] [PubMed]
26. García-Aguilar, J.; Hernandez de Anda, E.; Sirivongs, P.; Lee, S.H.; Madoff, R.D.; Rothenberger, D.A. A pathologic complete response to preoperative chemoradiation is associated with lower local recurrence and improved survival in rectal cancer patients treated by mesorectal excision. *Dis. Colon Rectum.* **2003**, *46*, 298–304. [CrossRef] [PubMed]
27. Maas, M.; Nelemans, P.J.; Valentini, V.; Das, P.; Rödel, C.; Kuo, L.-J.; A Calvo, F.; García-Aguilar, J.; Glynne-Jones, R.; Haustermans, K.; et al. Long-term outcome in patients with a pathological complete response after chemoradiation for rectal cancer: A pooled analysis of individual patient data. *Lancet Oncol.* **2010**, *11*, 835–844. [CrossRef]
28. Arredondo, J.; Baixauli, J.; Pastor, C.; Chopitea, A.; Sola, J.J.; González, I.; A-Cienfuegos, J.; Martínez, P.; Rodriguez, J.; Hernández-Lizoain, J.L. Mid-term oncologic outcome of a novel approach for locally advanced colon cancer with neoadjuvant chemotherapy and surgery. *Clin. Transl. Oncol.* **2017**, *19*, 379–385. [CrossRef]
29. Zhou, H.; Department of Colorectal Surgery; Song, Y.; Jiang, J.; Niu, H.; Zhao, H.; Liang, J.; Su, H.; Wang, Z.; Zhou, Z.; et al. A pilot phase II study of neoadjuvant triplet chemotherapy regimen in patients with locally advanced resectable colon cancer. *Chin. J. Cancer Res.* **2016**, *28*, 598–605. [CrossRef]
30. Petrelli, F.; Trevisan, F.; Cabiddu, M.; Sgroi, G.; Bruschieri, L.; Rausa, E.; Ghidini, M.; Turati, L. Total Neoadjuvant Therapy in Rectal Cancer: A Systematic Review and Meta-analysis of Treatment Outcomes. *Ann. Surg.* **2020**, *271*, 440–448. [CrossRef]
31. Cheong, C.K.; Nistala KR, Y.; Ng, C.H.; Syn, N.; Chang HS, Y.; Sundar, R.; Yang, S.Y.; Chong, C.S. Neoadjuvant therapy in locally advanced colon cancer: A meta-analysis and systematic review. *J. Gastrointest. Oncol.* **2020**, *11*, 847–857. [CrossRef] [PubMed]

medicina

MDPI

Article

Sensitivities and Dependencies of *BRAF* Mutant Colorectal Cancer Cell Lines with or without *PIK3CA* Mutations for Discovery of Vulnerabilities with Therapeutic Potential

Ioannis A. Voutsadakis [1,2]

[1] Algoma District Cancer Program, Sault Area Hospital, 750 Great Northern Road, Sault Ste. Marie, ON P6B 0A8, Canada; ivoutsadakis@yahoo.com or ivoutsadakis@nosm.ca

[2] Section of Internal Medicine, Division of Clinical Sciences, Northern Ontario School of Medicine, Sudbury, ON P3E 2C6, Canada

Abstract: *Background:* Colorectal cancer represents a common malignancy and remains incurable in the metastatic stage. Identification of molecular alterations that are present in colorectal cancer has led to the introduction of targeted therapies that improve outcomes. *BRAF* and *PIK3CA* mutations are observed in a subset of colorectal cancers. Colorectal cancers bearing *BRAF* mutations may be treated with specific BRAF inhibitors. These drugs benefit patients with *BRAF* mutant colorectal cancers but responses are rather brief, and progression is the rule. In contrast, no PI3K inhibitors have proven successful yet in the disease. Thus, new treatments to supplement the currently available drugs would be welcome to further improve survival. *Methods:* Profiled colorectal cancer cell lines from the Cancer Cell Line Encyclopedia (CCLE) were examined for *BRAF* and *PIK3CA* mutations and were interrogated for molecular characteristics and concomitant alterations that mirror clinical sample alterations. The Genomics of Drug Sensitivity in Cancer (GDSC) project was used for determination of drug sensitivities of *BRAF* mutated colorectal cell lines with or without concomitant *PIK3CA* mutations. The Cancer Dependency Map project served as the basis for identification of molecular dependencies and vulnerabilities in these cell lines. *Results:* CCLE includes 84 colorectal cancer cell lines, which recapitulate the molecular landscape of colorectal cancer. Of these, 23 and 24 cell lines possess *BRAF* and *PIK3CA* mutations, respectively. Seven *BRAF* mutant cell lines have V600E mutations and 14 *PIK3CA* mutant cell lines have hotspot helical or kinase domain mutations. V600E *BRAF* mutant cell lines with or without hotspot *PIK3CA* mutations are heterogeneous in their MSI status and mimic colorectal cancer tissues in other prevalent abnormalities including *APC* and *TP53* mutations. Essential genes for survival include *CTNNB1*, *WRN*, and pyrimidine metabolism enzyme *CAD*. Besides *BRAF* mutations, BRAF inhibitor sensitivity in colorectal cancer cell lines is conferred by *SACS* mutations and *PRKN* locus loss. *Conclusions:* Colorectal cancer cell lines bearing the frequent *BRAF* and *PIK3CA* mutations present many alterations of the parental cancer tissue. Described vulnerabilities represent leads for therapeutic exploration in colorectal cancers with the corresponding alterations.

Keywords: colon cancer; cell line models; dependencies; targeted therapy; signal transduction

check for updates

Citation: Voutsadakis, I.A. Sensitivities and Dependencies of *BRAF* Mutant Colorectal Cancer Cell Lines with or without *PIK3CA* Mutations for Discovery of Vulnerabilities with Therapeutic Potential. *Medicina* **2022**, *58*, 1498. https://doi.org/10.3390/medicina58101498

Academic Editors: Maria Rosaria De Miglio and Antonio M Scanu

Received: 30 August 2022
Accepted: 18 October 2022
Published: 21 October 2022

1. Introduction

Colorectal cancer is the most prevalent gastrointestinal carcinoma and a major cause of cancer morbidity and mortality. An estimated 150,000 people will be diagnosed with colorectal cancer in 2022 in the United States alone and over 50,000 patients will die from the disease [1]. It represents the third leading cause of mortality from cancer in both men (after lung and prostate cancers) and women (behind lung and breast cancers). About 20% of cases are diagnosed in a metastatic stage and a significant percentage of initially stage II and stage III patients will have a metastatic relapse [2]. Metastatic colorectal cancer remains most often an incurable disease, despite progress in systemic and local therapies that have improved outcomes [3]. The elucidation of the molecular pathogenesis of colorectal cancer has resulted in introduction of targeted therapies that have improved survival of selected patients [4–7]. These include anti-EGFR monoclonal antibodies for *KRAS* wild type disease, combinations of anti-EGFR monoclonal antibodies with BRAF inhibitors for *BRAF* mutant cancers, anti-HER2 therapies for HER2 altered cancers and immune checkpoint inhibitors for microsatellite instability (MSI) high cancers. Other targeted treatments addressing small defined sub-sets of colorectal cancers include NTRK inhibitors for colorectal cancers with NTRK fusions and specific KRAS G12C inhibitors for cancers with this KRAS substitution [8,9]. Novel therapeutics based on combinations of targeted therapies are intensely investigated with the hope that several will enter the clinic in the near future [10,11].

BRAF mutations are observed in 5% to 15% of colorectal cancers and are associated with aggressive disease [12,13]. Colorectal cancers with mutations in *BRAF* tend to be of high grade and occur more often in the right colon [14]. The most common mutations in *BRAF* occur at amino-acid V600 position of the protein and substitute the normal valine at this position with glutamic acid (V600E). *BRAF* V600E mutations and other rarer substitutions at this codon location (V600K, V600D, V600M, and V600R) are categorized as class I *BRAF* mutations. These substitutions result in potent kinase activation that is independent of upstream signals from KRAS [15,16]. Mutations of *BRAF* in other codons, including the neighboring L597 and K601 positions lead to a protein that retains the requirement for homo-dimerization to signal downstream. These mutations that are classified as class II, as well as class III mutations, that require KRAS input for sustained signaling, are rare [14,15].

Mutations in the gene encoding for the alpha catalytic subunit of kinase PI3K, *PIK3CA*, are the most common colorectal cancer mutations in the PI3K/AKT/mTOR signal transduction pathway and are present in 20% to 25% of colorectal cancers [17–20]. *PIK3CA* point mutations are more diverse than *BRAF* mutations, although about half of the cases concern codons E542, E545, and Q546 of the helical domain and codon H1047 of the kinase domain. Colorectal cancers with *PIK3CA* mutations are more often arising in the right colon and present with a higher mutation count than cancers without *PIK3CA* mutations [20]. In contrast to the mutual exclusivity of mutations in oncogenes *KRAS* and *BRAF*, cancers with *PIK3CA* mutations have often concomitant mutations in either of these genes of the KRAS/BRAF/MEK/ERK pathway.

This investigation examines colorectal cancer cell lines bearing *BRAF* mutations with concomitant *PIK3CA* mutations and compares them to *BRAF* mutant cell lines without *PIK3CA* mutations in regard to genomic characteristics such as ploidy, MSI status, and coexisting molecular alterations. The sensitivity of these cell lines to drugs inhibiting the mutated pathways and to other inhibitors is also interrogated. The ultimate goal is to discover new therapeutic opportunities beyond the currently available BRAF inhibitors, which are currently the only approved drugs, in combination with anti-EGFR therapies, for colorectal cancers with V600E mutations.

2. Methods

Cancer cell lines included in the current investigation constitute part of the Cancer Cell Line Encyclopedia (CCLE) collection [21]. The cBioportal Genomics Portal platform was used to identify colorectal cancer cell lines with *BRAF* mutations with or without concomitant *PIK3CA* mutations in CCLE [22]. cBioportal (http://www.cbioportal.org accessed on 29 July 2022) is a user-friendly, open-access platform for genomic analysis of tumors and cancer cell lines [22]. Additionally, genomic data of colorectal cancer patients from The Cancer Genome Atlas (TCGA) study cohort [17] were analyzed using cBioportal. The CCLE project employs whole-exome sequencing to discover mutations, copy number alterations, and fusions in cell lines from various types of cancer [21]. Analysis of copy number alterations in the CCLE project was performed with the GISTIC (Genomic Identification of Significant Targets in Cancer) algorithm, in which a score of 2 or above denotes putative amplification of a gene [23]. RNA expression was normalized with the RSEM algorithm and results were presented as the Log RNA sequences in Reads per Kilobase Million (RPKM) [24].

The functional assessment of mutations observed in cell lines of interest was performed with the help of OncoKB. OncoKB knowledgebase is a database of cancer-related genes and characterizes these genes as oncogenes or tumor suppressor genes [25]. On some occasions, genes are included in OncoKB as cancer associated but they are not annotated as oncogenes or tumor suppressors.

The Genomics of Drug Sensitivity in Cancer (GDSC) dataset (www.cancerrxgene.org accessed on 29 July 2022) was interrogated to obtain data on drug sensitivity of cell lines from colorectal cancer and other cancers with *BRAF* and *PIK3CA* mutations [26]. Two datasets, GDSC1 and GDSC2, are included within the GDSC project, differing in the experimental conditions used. GDSC1 experiments were performed between 2009 and 2015. These experiments used media alone in the negative control cell lines not exposed to drugs. The GDSC2 panel of experiments was performed more recently (after 2015) and employed media with vehicle (DMSO-dimethylsulfoxide) in the negative controls. Dependencies on specific genes of cell lines with *BRAF* and *PIK3CA* mutations were obtained from the Depmap portal that contains data from CRISPR arrays and RNA-interference (RNAi) arrays of included cell lines from CCLE [27,28]. CRISPR and RNAi arrays identify essential genes that are important for the survival of screened cell lines and, as a result, the knock-down of these essential genes has a significant effect in their survival and proliferation in vitro [29–31]. The two methodologies differ in the depth of suppression of assayed genes, with CRISPR knock out usually being stronger than the partial suppression obtained by RNA interference. As a result, the genes and dependencies discovered with the two methodologies are not completely overlapping. Data for CRISPR screening in DepMap are from project SCORE containing 323 cancer cell lines from various cancers and a library of 18,009 targeted genes [32]. Computational modelling of experiments in SCORE was initially performed with the CERES algorithm and later with the CHRONOS algorithm [33,34]. RNAi experiments were performed under the aegis of project Achilles using the DEMETER algorithm for analysis [30].

Statistical comparisons of categorical data were carried out using Fisher's exact test or the x^2 test. The Mann–Whitney U test was used to compare median values. All statistical comparisons were considered significant if $p < 0.05$.

All data presented in this paper are from experiments performed by the consortiums mentioned in the above methods section and are openly available in the public domain. No new laboratory experiments have been performed for this investigation.

3. Results

The colorectal cancer cohort of CCLE consisting of 84 cell lines contains 23 cell lines (27.4%) with *BRAF* mutations. Ten *BRAF* mutant cell lines contain classic V600E mutations, in three of them (OUMS23, MDST8 and HT-29) with additional non-canonical *BRAF* mutations (Table 1). Thirteen cell lines contain non-V600E mutations. In two of them, NCI-H508 and HT-55, mutations are oncogenic or potentially oncogenic (G596R and N581Y, respectively).

Seven *BRAF* V600E mutant cell lines are wild type for *PIK3CA*, while three cell lines with V600E mutations (SNU-C5, RKO and HT-29) as well as cell line NCI-H508, which has a pathogenic non-V600 mutation at position G596, have concomitant pathogenic mutations in *PIK3CA* (Table 1). Five of the seven cell lines with V600E *BRAF* mutations and no *PIK3CA* mutations are MSS, possess a lower mutation count, are hyper-diploid and have a high Fraction of Genome Altered (FGA) (Table 2). The two V600E *BRAF* mutant/*PIK3CA* wild type colorectal cancer cell lines, LS411N and CL34, that are MSI high have consistently a high mutation count. The two cell lines with concomitant *BRAF* V600E and *PIK3CA* H1047R mutations, SNU-C5 and RKO, are MSI high, have a high mutation count, are diploid and have a low FGA (Table 2). The two other cell lines with concomitant mutations, NCI-H508 and HT-29, have non-canonical pathogenic mutations in either *BRAF* (NCI-H508) or in *PIK3CA* (HT-29) and they are both MSS, have lower mutation counts, are hyper-diploid and have a high FGA.

Table 1. *BRAF* mutated colorectal cancer cell lines and their specific *BRAF* mutations and concomitant *PIK3CA* mutations. Data are from the Cancer Cell Line Encyclopedia (CCLE). WT: wild type.

Cell Line	*BRAF*	*PIK3CA*
BRAF V600E mutations		
COLO205	V600E	WT
COLO201	V600E	WT
LS411N	V600E	WT
SW1417	V600E	WT
CL34	V600E	WT
MDST8	V600E, V600K, V600M	WT
OUMS23	V600E, X287_splice	WT
SNU-C5	V600E	H1047R
RKO	V600E	H1047R
HT-29	V600E, T119S	P449T
BRAF non-V600E pathogenic mutations		
HT-55	N581Y	WT
NCI-H508	G596R	E545K
BRAF mutations of unknown significance		
HT115	R354Q	R88Q, E321D, R770Q
SNU-C4	D22N	E545G, V71I
CCK81	S273N, R506G	C420R, C472Y
LS513	E204L, E204V, E204*	WT
GP2D	T529A	H1047L
SNU1040	V120I, S76P	L632*
SNU407	R726C	H1047R
SNU503	D22N	WT
LS180	D211G	H1047R
KM12	A712T, A404Cfs*9	WT
GP5D	T529A	H1047L

Table 2. Characteristics of colorectal cancer cell lines with *BRAF* V600E mutations without and with concomitant *PIK3CA* mutations. Cell line NCI-H508 has a *BRAF* G596R pathogenic mutation instead of *BRAF* V600E mutation. Cell lines without an asterisk are without *PIK3CA* mutations and are presented first. Cell lines with an asterisk in the bottom lines of the table are those with concomitant *PIK3CA* mutations.

Cell Line	DepMap ID	Mutation Count	FGA	Ploidy	MSI Status
COLO205	ACH-001039	307	0.44	3.2	MSS
OUMS23	ACH-000296	340	0.50	2.5	MSS
COLO201	ACH-000253	255	0.38	2.96	MSS
MDST8	ACH-000935	776	0.55	3.82	MSS
LS411N	ACH-000985	5442	0.28	3.30	MSI
SW1417	ACH-000236	248	0.56	3.01	MSS
CL34	ACH-000895	1280	0.14	1.95	MSI
SNUC5 *	ACH-000970	2990	0.09	2.0	MSI
RKO *	ACH-000943	3424	0.14	2.1	MSI
NCI-H508 *	ACH-000360	318	0.48	4.6	MSS
HT-29 *	ACH-000552	416	0.43	3.04	MSS

Regarding concomitant cancer-associated mutations in V600E *BRAF* mutant/*PIK3CA* wild type colorectal cancer cell lines all seven cell lines have oncogenic mutations in *APC* and four have also oncogenic mutations in *TP53* (Table 3). No cell lines have *KRAS* mutations, which tend to be mutually exclusive with *BRAF* mutations. Recurrent oncogenic deletions include the loci of dual specificity phosphatase *DUSP22*, which is present in 4 cell lines and deletions in *SMAD4* and *SMAD2*, which are present in 3 and 2 cell lines, respectively (Table 3). Only two of the four cell lines with oncogenic mutations in both *BRAF* and *PIK3CA* have concomitant *APC* mutations and three of the four have also *TP53* mutations (Table 3). Recurrent amplifications are observed in *MYC* and *AGO2* that are both located at chromosome arm 8q and are present in cell lines RKO and HT-29. These cell lines and the cell line NCI-H508 also possess deletions of *PRKN*, encoding for ubiquitin ligase parkin, which is the only recurrent deletions in *BRAF*/*PIK3CA* double mutant colorectal cancer cell lines. HT-29 is the only double mutant cell line possessing the recurrent deletion of *DUSP22*, observed in cell lines with V600E *BRAF* mutations and wild type *PIK3CA* (Table 3).

Table 3. Molecular alterations in colorectal cancer cell lines with *BRAF* V600E mutations without and with concomitant *PIK3CA* mutations. +: presence of oncogenic mutation. Cell lines with an asterisk are those with concomitant *PIK3CA* mutations.

Cell Line	APC	TP53	KRAS	SMAD4	ATM	FBXW7	Other Mutations	Amplifications	Deletions
COLO205	+							CCND3	CDC73, DUSP22, SMAD4
OUMS23	+	+					TBX3	AURKA, YES1	PTEN, MAP2K4, CDC73, FAT1, SMAD4, SMAD2, BMPR1A
COLO201	+						EPHA7, BACH2		
MDST8	+						CDKN2A		DUSP22, HLA-A, PAX5, BCL11B
LS411N	+	+				+	BARD1, BRIP1, PTEN, ARID1A, RNF43, MLH1, KMT2A, KMT2A, KMT2D	FGFR1	HLA-A, SMAD4, SMAD2, PMAIP1
SW1417	+	+					RTEL	MET, AURKA, BRAF, SRC, BCL2L1, EZH2, RHEB, DNMT3B	IKZF1, FLCN, NKX3-1, DUSP22, PIK3R1, PPP2R2A
CL34	+	+					TSC1, HLA-B, TCF7L2, PPM1D, PARP1, TP53BP1, CREBBP, TGFBR2, AMER1, ATRX, AXIN2, SOX9, ARID4B, MAP2K4		DUSP22, FOXP1, H1-3, JARID2,
SNUC5 *	+	+			+	+	ERCC2, ARID1A, CTCF, ARID2, RNF43, PPM1D, DNMT3A, ZFHX3, CREBBP, CYLD, EP300, LATS1, KMT2C, FANCC, KMT2B, ARID4A, NCOR1, ASXL2, KMT2D, BCORL1, CD58, ELF3, MED12, EP400	AR	FLCN, PTPRT
RKO *							BRCA2, ARID1A, NF1, STAT3, KMT2A, B2M, BCORL1, EP300, RNF43, JARID2, NCOR1, PARP1, TET1, TP53BP1, CREBBP, FANCA, NOTCH3, NF2, NSD1, PTPRD, MSH6, FAT1, GATA3, SOX9	UBR5, AGO2, MYC	PTPRD, PRKN, PAX5, FOXA1, EP300, INHA
NCI-H508 *		+					CDKN2A, SPOP, CREBBP	PIK3CA, BCL2L1, BCL6, DNMT3B, CDK8	PRKN, INPP4B, TCF7L2
HT-29 *	+	+		+			CASP8, SLFN11	KIT, AGO2, MYC, CDK8	NKX3-1, DUSP22, PRKN, ESCO2, PPP2R2A, EPHA3

Vulnerabilities of *BRAF* mutant cell lines with or without *PIK3CA* mutations were explored with interrogation of RNAi libraries for determination of preferentially essential genes and with CRISPR mediated knock out arrays (Table 4). Recurrent genes that are observed to be essential for survival in more than one *BRAF* mutant cell lines include *CTNNB1*, encoding for β-catenin, *WRN*, encoding for Warner syndrome ATP-dependent helicase, *ALYREF* which encodes for a chaperone of basal region leucine zipper (bZIP) proteins, and peptidylprolyl isomerase E (PPIE). These recurrent essential genes are in the top list of preferentially essential genes in one or more of the four cell lines with *BRAF* and *PIK3CA* mutations (Table 4). In addition, the gene encoding for CAD, an enzyme of the pyrimidine biosynthesis pathway induced by MAPK cascade, is a preferentially essential gene in two of four *BRAF* and *PIK3CA* mutant cell lines.

Table 4. Top dependencies of *BRAF* V600E mutant/*PIK3CA* wild type and *BRAF* V600E mutant/*PIK3CA* mutant colorectal cancer cell lines, as determined by RNAi and CRISPR knock-out. RNAi experiments are from project Achilles and CRISPR experiments are from project SCORE and CHRONOS. NA: not available.

Cell Line	Top 10 Preferentially Essential Genes RNAi	Top CRISPR KO Genes
COLO205	NA	YRDC, ADSL, MMS22L, UMPS, TRNT1 (SCORE)
OUMS23	GSPT1, ALYREF, BUB3, BUB1B, RPL13, PHB, QARS1, SERPINA5, MAD2L1,ZRSR2	SLC25A37, SCAP, ATP6V1A, RGP1, SOD2, CHAF1B, CIAO2B, CHAF1A, ATP6VOB, DHX9 (CHRONOS)
COLO201	BRAF, MAP2K1, MYBL2, BCL2L1, CTNNB1, MAPK1, SOX9, CHD4, TCF7L2, PSMD2	PSMG4, MYB, CLNS1A. HSPA8, RUVBL1 (SCORE)CTNNB1, DUSP4, HSPA8, WASF2, SOX9, SLC1A5, NIBAN2, BRAF, ASCL2, IQGAP1 (CHRONOS)
MDST8	NA	HSPD1, USP17L5, MDM2, YRDC, CYCS (SCORE)
LS411N	CTNNB1, WRN, DDX39A, BCL2L1, ALYREF, DHX9, JPT2, PCNA, PPIE	TINF2, RNPC3, NXT1, HSPA9, NUP85 (SCORE)NXT1, SLC7A1, CTNNB1, DDX39A, WRN, BCL2L1, INTS6, ADSL, SYNCRIP, SNAP23 (CHRONOS)
SW1417	CDC40, PPIE, KHSRP, EFCAB8, PPP2CA, PPWD1, EIF4A3, MED11, OR56B1, CAPZB	NA
CL34	CTNNB1, WRN, ZNF432, SCAP, CWC22, NUP214, TTC1, RPA3, SAP130, PHB	NA
SNUC5	NA	CDCA8, YRDC, WDR82, FAU, BUD31 (SCORE)WRN, NAMPT, TRPM7, RFK, ADSL, PELO, NDE1, MTHFD1, PPIE, RAB1 (CHRONOS)
RKO	OGDH, WRN, ALDH18A1, URI1, RPL22L1, CAD, TTC7A, CD3EAP, SDHD, SDHC	CCT4, DYNLRB1, UBE2M, FAU, RPP21 (SCORE)ATPV0E1, WRN, CREBBP, TTC7A, CAD, SLC5A3, MTCH2, MTX2, UMPS, FAM126B (CHRONOS)
NCI-H508	MYBL2, SLC22A20P, TYMS, ANKRD2019P, SKP1, PSMA3, CTNNB1, YAP1, EFCAB8, EGFR	NA
HT-29	RAB6A, GINS2, APC, AHCTF1, BRAF, COP1, CAD, PFAS, SUMO2	DYNLRB1, THAP1, MYC, PPP2CA, INTS6 (SCORE)PTDSS1, INTS6, RIC1, SCD, SCAP, MBTPS2, NDE1, STX4, RAB10, HNF1B (CHRONOS)

Five of the seven cell lines with *BRAF* mutations and without *PIK3CA* mutations (COLO205, MDST8, LS411N, SW1417 and CL34) have been assayed for drug sensitivities in GDSC (Table 5). Top drug sensitivities displayed by cell lines COLO205 and CL34 are to BRAF inhibitors, inhibitors of downstream MEK kinases and inhibitors of upstream receptor tyrosine kinases. LS411N cell line displays sensitivity to drugs of the pathway as well as to other kinases and the dihydrofolate reductase inhibitor pyrimethamine. In contrast, no inhibitors of BRAF or the receptor tyrosine kinase/KRAS/BRAF/MAPK pathway are among the top sensitivities of cell lines MDST8 and SW1417. Top sensitivities of these two cell lines include drugs involved in lipid metabolism and apoptosis inhibitors (Table 5). Cell lines with mutations in both *BRAF* and *PIK3CA* display sensitivities to several inhibitors of the receptor tyrosine kinase/KRAS/BRAF/MAPK pathway and PI3K/AKT cascade. Two of the four *BRAF*/*PIK3CA* double mutated cell lines, SNUC5 and RKO present additional sensitivities to the clinically used antimetabolite methotrexate, the WEE1 kinase inhibitor MK-1775, the mitotic kinases AURKA and AURKB inhibitor ZM447439 and the epigenetic modifier, BET bromodomain inhibitor JQ1. Compared with cell lines not bearing mutations in *BRAF* and *PIK3CA*, colorectal cancer cell lines with *BRAF* mutations with or without *PIK3CA* mutations show heterogeneous up-regulation in the mRNA expression of genes that are targets of the BRAF/MEK/ERK pathway. These include phosphatases DUSP5,

DUSP6, AP-1 transcription factor component FOS, and apoptosis inhibitors survivin (also known as BIRC5—that is, baculoviral IAP repeat containing 5) and MCL1 (Figure 1). However, the robustness of pathway upregulation as suggested by the upregulation of these genes does not correlate with sensitivity to BRAF inhibitors. For example, cell lines SW1417 and MDST8, which display upregulation of pathway target genes, show no BRAF or other pathway inhibitors among their top inhibiting drugs (Table 5).

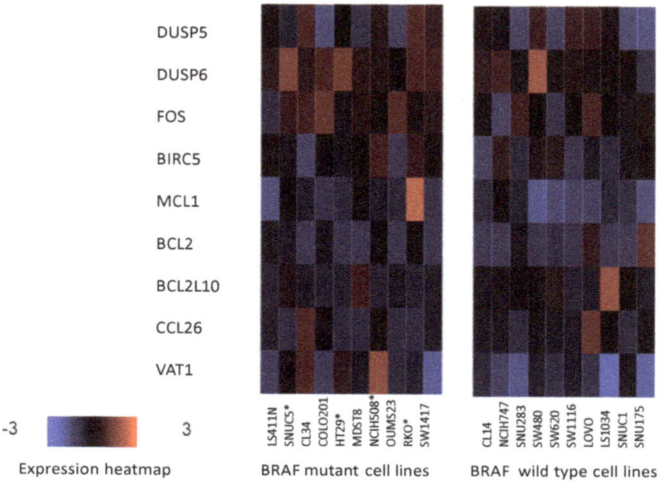

Figure 1. mRNA expression of genes targeted by the BRAF/MEK/ERK pathway (DUSP5, DUSP6, FOS, BIRC5, and MCL1) and genes not directly targeted by the BRAF/MEK/ERK pathway (BCL2, BCL2L10, CCL26 and VAT1) as controls in representative colorectal cancer cell lines with (left panel) and without (right panel) mutations in *BRAF*. BRAF mutated cell lines with coexisting PIK3CA mutations are shown with an asterisk.

GDSC includes five specific BRAF inhibitors among the panel of assayed drugs. Recurrent molecular characteristics of the colorectal cancer cell lines panel that confer sensitivity to specific BRAF inhibitors include, as expected, *BRAF* mutations conferring sensitivity to 4 of the 5 inhibitors (Table 6). In addition, the presence of *KRAS* mutations confer resistance to 3 of the 5 BRAF inhibitors, as they tend to be mutually exclusive with *BRAF* mutations and segregate with *BRAF* wild type cell lines. Another genomic feature that is present recurrently among the abnormalities conferring BRAF inhibitor sensitivity in colorectal cancer cell lines is mutations in SACS, a gene encoding for sacsin, a chaperone protein. The most common copy number alteration that confers resistance to 3 of the 5 BRAF inhibitors is a loss at chromosome 6q26, a locus containing gene *PRKN*, encoding for E3 ubiquitin ligase parkin (feature cnaCOREAD24). Loss of *PRKN* is a feature of some *BRAF* mutant cell lines, as mentioned above, and it is also, rarely, encountered in *BRAF* mutant colorectal cancers. Thus, resistance to BRAF inhibitors associated with concomitant loss of *PRKN* may be of clinical significance. Interestingly, *PIK3CA* mutations do not feature among the molecular abnormalities conferring resistance to specific BRAF inhibitors in colorectal cancer cell lines. The only BRAF specific inhibitor that is not significantly more effective in *BRAF* mutant cell lines is HG6-64-1, which displays a separate private panel of mutations conferring resistance, not observed in other BRAF inhibitors. These include EGFR mutations and mutations in kinase ATM (Table 6).

Table 5. Drug sensitivities of *PIK3CA* wild type/*BRAF* V600E mutant cell lines. Data are from the Genomics of Drug Sensitivity in Cancer (GDSC).

Cell Line	Drug	Target	IC50	Z Score	Source
COLO205	SB590885	BRAF	0.18	−4.13	GDSC1
	PLX-4720	BRAF	0.19	−4.04	GDSC1
	Selumetinib	MEK1, MEK2	0.08	−3.39	GDSC1
	BMS-754807	IGF1R, IR	0.01	−3.17	GDSC1
	Lisitinib	IGF1R	0.09	−3.15	GDSC2
MDST8	CAY10566	Steroyl-CoA Desaturase	0.07	−4.20	GDSC1
	SGC0946	DOT1L	0.99	−3.03	GDSC1
	CCT007093	PPM1D	7.7	−2.76	GDSC1
	UNC1215	L3MBTL3	2.37	−2.61	GDSC1
	(5Z)-7-Oxozeanol	TAK1	0.04	−2.57	GDSC1
LS411N	Pyrimethamine	Dihydrofolate reductase	0.72	−2.82	GDSC1
	VX11e	ERK2	0.56	−2.33	GDSC1
	AZ628	BRAF	0.11	−2.13	GDSC1
	Alectinib	ALK	3.98	−2.12	GDSC1
	GNF-2	BCR-ABL	2.20	−2.01	GDSC1
SW1417	WEHI-539	BCL-XL	0.33	−2.48	GDSC2
	Sphingosine kinase 1 inhibitor II	Sphingosine kinase	10.2	−2.05	GDSC1
	CHIR-99021	GSK3A, GSK3B	3.07	−1.99	GDSC1
	Navitoclax	BCL2, BCL-XL, BCL-W	0.28	−1.61	GDSC2
	SN-38	TOP1	0.00	−1.43	GDSC1
CL34	Trametinib	MEK1, MEK2	0.00	−2.92	GDSC2
	Dabrafenib	BRAF	0.16	−2.71	GDSC2
	SCH772984	ERK1, ERK2	0.06	−2.61	GDSC2
	Selumetinib	MEK1, MEK2	0.06	−2.34	GDSC1
	PLX-4720	BRAF	2.77	−2.01	GDSC1
SNUC5	Methotrexate	Antimetabolite	0.04	−1.52	GDSC1
	PD0325901	MEK1, MEK2	0.04	−1.20	GDSC1
	Bosutinib	SRC, ABL	1.16	−1.18	GDSC1
	PLX-4720	BRAF	13.55	−1.13	GDSC1
	MK-1775	WEE1	0.48	−1.12	GDSC1
RKO	KIN-001	GSK3B	13.4	−2.7	GDSC1
	Selumetinib	MEK1/2	0.29	−2.49	GDSC1
	AZ628	BRAF	0.06	−2.47	GDSC1
	ZM447439	AURKA/B	0.58	−2.19	GDSC1
	JQ1	BRD2/3/4	0.05	−2.13	GDSC1
NCI-H508	Afatinib	ERBB2, EGFR	0.04	−2.81	GDSC1
	Afatinib	ERBB2, EGFR	0.07	−2.71	GDSC2
	Gefitinib	EGFR	0.23	−2.12	GDSC1
	Pictilisib	PI3K (class 1)	0.18	−2.00	GDSC1
	MK-2206	AKT1, AKT2	0.87	−1.97	GDSC2
HT-29	ERK_6604	ERK1, ERK2	0.62	−2.20	GDSC2
	BMS-754807	IGF1R, IR	0.05	−2.17	GDSC1
	Linsitinib	IGF1R	0.42	−2.08	GDSC1
	Refametinib	MEK1, MEK2	0.13	−1.98	GDSC1
	AS605240	PI3Kgamma	1.04	−1.98	GDSC1

Table 6. Top molecular features with increased sensitivities to various BRAF inhibitors (statistically significant or approaching significance). Two non-specific RAF inhibitors (RAF 9304 and Sorafenib) are also shown. Data are from the Genomics of Drug Sensitivity in Cancer (GDSC).

Drug	Feature	IC50 Effect Size	p Value	Number of Altered Cell Lines	Dataset
AZ628	SACS mutation	−2.45	0.007	3	GDSC1
	cnaCOREAD19	−2.05	0.008	4	GDSC1
	BRAF mutation	−1.52	0.04	5	GDSC1
	KRAS mutation	−0.44	0.08	6	GDSC1
	FBXW7 mutation	−1.26	0.08	3	GDSC1
Dabrafenib	BRAF mutation	−2.24	2.21×10^{-7}	10	GDSC2
	KRAS mutation	0.85	0.006	24	GDSC2
	cnaCOREAD24	0.99	0.012	9	GDSC2
	KDM6A mutation	1.21	0.016	3	GDSC2
	cnaCOREAD55	0.89	0.02	10	GDSC2
	cnaCOREAD56	0.89	0.02	10	GDSC2
	SACS mutation	−0.82	0.031	10	GDSC2
HG6-64-1	ATM mutation	1.33	0.02	3	GDSC1
	SMARCA4 mutation	0.41	0.033	3	GDSC1
	EGFR mutation	0.39	0.037	3	GDSC1
	PBRM1 mutation	1.18	0.039	3	GDSC1
PLX-4720	BRAF mutation	−1.77	1.58×10^{-5}	10	GDSC2
	KRAS mutation	0.91	0.002	25	GDSC2
	SACS mutation	−1.03	0.009	10	GDSC2
	cnaCOREAD19	−0.83	0.013	18	GDSC2
	cnaCOREAD55	0.88	0.018	11	GDSC2
	cnaCOREAD56	0.88	0.018	11	GDSC2
	cnaCOREAD23	1.46	0.026	3	GDSC2
	cnaCOREAD53	1.46	0.026	3	GDSC2
	cnaCOREAD24	0.84	0.034	9	GDSC2
	BCOR mutation	−1.05	0.045	5	GDSC2
SB590885	BRAF mutation	−1.21	0.002	10	GDSC1
	cnaCOREAD24	1.16	0.004	9	GDSC1
	KRAS mutation	0.8	0.009	25	GDSC1
	cnaCOREAD12	1.65	0.012	3	GDSC1
	SACS mutation	−0.95	0.018	10	GDSC1
	cnaCOREAD56	0.83	0.029	11	GDSC1
	cnaCOREAD55	0.83	0.029	11	GDSC1
RAF 9304	cnaCOREAD63	1.25	0.009	6	GDSC1
(pan-RAF)	TP53 mutation	0.96	0.009	33	GDSC1
	ARID1B mutation	1.17	0.036	3	GDSC1
	PIK3R1 mutation	1	0.047	5	GDSC1
	BRAF mutation	−0.76	0.047	10	GDSC1
	NCOR1 mutation	−0.95	0.049	6	GDSC1
Sorafenib	cnaCOREAD47	−0.89	0.01	6	GDSC2
(PDGFR, c-KIT,	KDM6A mutation	1.05	0.01	6	GDSC2
VEGFR, RAF)	cnaCOREAD48	1.24	0.026	5	GDSC2
	CEP290 mutation	−1.09	0.029	5	GDSC2
	KRAS mutation	0.63	0.037	24	GDSC2
	cnaCOREAD14	1.19	0.049	3	GDSC2

In the pan-cancer analysis of cell lines with *BRAF* mutations, which is more statistically robust due to the number of cell lines assayed, pathway inhibitors (BRAF inhibitors:

Dabrafenib, PLX-4720, SB59088, MEK inhibitors: selumetinib, trametinib, refametinib, PD0325901, ERK inhibitors: ulixertinib, ERK2440, ERK6604, SCH772984, VX-11e) are significantly associated with sensitivity compared to cell lines without *BRAF* mutations. In addition, the inhibitor of NUAK1 and NUAK2 kinases WZ4003 is statistically significantly associated with sensitivity in BRAF mutant cell lines compared with *BRAF* wild type cell lines (IC50 effect size: -0.34, $p = 8.03 \times 10^{-5}$). Specifically for colorectal cancer cell lines, *BRAF* mutant cell lines display also greater sensitivity to inhibitor WZ4003 compared to *BRAF* wild type colorectal cancer cell lines (mean IC50: 63.7 μM versus 132 μM), although, due to smaller numbers, this difference did not reach statistical significance ($p = 0.08$).

4. Discussion

BRAF is an oncogenic serine/threonine kinase, which is mutated in various cancers, most commonly in melanoma, thyroid carcinomas, hairy cell leukemia, lung cancers, and colorectal cancers [35]. The gene encoding for the kinase is located on the human chromosome locus 7q34. BRAF is activated by KRAS downstream of growth factor receptors and activates the Mitogen Activated Protein Kinase (MAPK)/Extracellular signal-Regulated Kinase (ERK) signaling cascade promoting cell proliferation. The importance of this pathway in cancer is highlighted by the fact that KRAS is the most frequently mutated oncogene across cancer types [36]. In parallel with the KRAS/BRAF/MAPK/ERK pathway, and also activated by growth factor receptors, the PI3K/AKT/mTOR cascade plays an important role in carcinogenesis through inhibition of apoptosis, cell growth promotion and oncogene activation [37]. *PIK3CA*, the gene encoding for the catalytic alpha sub-unit of kinase PI3K is often mutated in prevalent cancers such as breast cancer and colorectal adenocarcinomas. In colorectal cancer, *PIK3CA* is mutated in 20% to 25% of cases and is the second most commonly mutated oncogene after *KRAS* [17]. *BRAF* mutated colorectal cancers are less prevalent, representing 5% to 15% of all colorectal cancers. Most of *BRAF* mutations are located at amino acid position V600, substituting glutamic acid for valine that is normally at this position in the wild type protein (V600E substitution). Substitutions at position V600 render the protein independent from KRAS and result in robust kinase-mediated activation of MAPK cascade, without the physiologic input from growth factors [38]. Other less common *BRAF* mutations produce a protein with lower kinase activity or even a kinase-dead protein that can still activate down-stream signaling through interaction with the homologous CRAF kinase [15]. Canonical V600E BRAF mutations are mutually exclusive with *KRAS* mutations. In contrast, *PIK3CA* mutations are encountered in colorectal cancers with either *KRAS* or *BRAF* mutations with an equal or higher prevalence than in cancers with wild type *KRAS* and *BRAF*.

BRAF mutations are targeted currently in colorectal cancer in the clinic at the second line metastatic setting with a regimen that combines BRAF inhibitors and anti-EGFR monoclonal antibodies. This combination has provided superior efficacy and survival outcomes compared with chemotherapy, with a modest improvement of 3 months in Overall Survival (OS) [39]. In contrast, no therapies targeting *PIK3CA* mutated colorectal cancers have been approved for clinical use. Combinations of BRAF inhibitors with PI3K inhibitors have not been studied in a systematic manner in colorectal cancer, but few available retrospective data suggest that parallel inhibition of the two mutated oncogenes may provide a synergistic effect in double mutant cancers [40]. Unveiling vulnerabilities of colorectal cancers with *BRAF* mutations with and without concomitant *PIK3CA* mutations may provide new opportunities for targeted treatments.

The current investigation examines a panel of colorectal cancer cell lines with *BRAF* mutations with or without concomitant mutations in *PIK3CA* from the CCLE for drug sensitivities and molecular dependencies. Mutations in *PIK3CA* are the most frequent mutations in the receptor tyrosine kinase-initiated pathways in colorectal cancers with *BRAF* mutations, as the even more frequent *KRAS* mutations are mutually exclusive with *BRAF* mutations. Colorectal cancer cell line models recapitulate the presence of *BRAF* and *PIK3CA* mutations as encountered in clinical colorectal cancer samples, and also duplicate the frequent presence of MSI in these cases [41]. Mutations in tumor suppressors *APC* and *TP53* are often present in *BRAF* mutant colorectal cancer cell lines, similar to clinical samples. Cell lines with *BRAF* mutations and wild type *PIK3CA* possess also deletions of signal transducers of TGFβ pathway SMAD4 and SMAD2 and of phosphatase DUSP22. The genes of these proteins are rarely deleted in clinical colorectal cancer, but they are more commonly mutated. For example, in TCGA cohort, *SMAD4* mutations are observed in 16.1% of cases with *BRAF* mutations, *SMAD2* mutations are observed in 6.5% of cases with *BRAF* mutations and *DUSP22* mutations are encountered in 9.7% of patients with *BRAF* mutations [17]. The presence of mutations or deletions of these genes suggest that decreased availability and function of the resulting proteins may be essential for *BRAF* mutant cancers both in vitro and in vivo. The TGFβ signaling pathway and tumor suppressor *SMAD4* mutations have been implicated in the serrated colon carcinogenesis pathway commonly resulting from *BRAF* mutations [42]. In addition, inhibitors of the TGFβ receptor TGFBR1 prevented the development of resistance to BRAF inhibitor vemurafenib in *BRAF* mutant melanoma cells [43]. Thus, inhibitors of the TGFβ pathway, should they become clinically available, could be candidates for combination therapies in *BRAF* mutated colorectal cancers. Phosphatase DUSP22 (also called JKAP- c-JUN N-terminal Kinase Associated phosphatase) is a regulator of the MAPK pathway, and as a result, it may modulate the effect of *BRAF* mutations in the pathway output [44]. DUSP22 showed lower mRNA expression in colorectal cancer tissues compared to adjacent normal colonic mucosa [45]. In this study that included 92 patients, patients with metastatic colorectal cancer and low expression of DUSP22 had a trend towards worse survival, although not statistically significant [45].

The analysis of molecular features associated with sensitivity or resistance to BRAF specific inhibitors reveals that, besides *BRAF* mutations and *KRAS* mutations that are associated with sensitivity and resistance to the drugs, respectively, no other abnormalities of the pathway affect sensitivity to these drugs in a consistent manner, in vitro. Unrelated molecular alterations associated with sensitization of colorectal cancer cell lines to BRAF inhibitors included mutations in *SACS*, encoding for chaperone protein sacsin and deletions at the locus of parkin. Sacsin is a large protein with chaperone function in the nervous system and loss of function mutations are associated with the degenerative disorder autosomal recessive spastic ataxia of Charlevoix-Saguenay [46]. Cells with sacsin loss of function have defective mitochondrial dynamics and increased oxidative stress. Mutations in *SACS* have not been previously linked with colorectal cancer. The protein consists of 4579 amino acids and is mutated in 12.5% of colorectal cancers of the TCGA cohort with mutations distributed equally across the length of the protein [17]. It is also mutated in 33.9% of colorectal cancers with *BRAF* mutations and in 19% of cancers with *PIK3CA* mutations. Among colorectal cancers classified as MSI high or with proofreading polymerase epsilon mutations, *SACS* mutations are present in 42.5% of cases, suggesting that these mutations are associated with high TMB and may be passenger [47]. Alternatively, an oncogenic role of sacsin mutations in colorectal cancer is also possible based on its function in oxidative stress and deserves to be formally confirmed or excluded.

Concomitant mutations in *APC* that are observed in most cell lines with *BRAF* mutations with or without *PIK3CA* mutations, as well as the fact that *CTNNB1* gene, encoding for β-catenin, is a recurrent preferential essential gene in these cell lines suggest that *BRAF* mutated colorectal cancers remain dependent on the activity of WNT/APC/β-catenin pathway [48,49]. Two other recurrent preferentially essential genes in *BRAF* mutated cell lines are *WRN*, encoding for Werner helicase and *CAD* (carbamoyl-phosphate synthetase 2, aspartate transcarbamylase and dihydroorotase), encoding for a protein with trifunctional enzyme activity implicated in the de novo pyrimidine nucleotide biosynthesis. WRN helicase is involved in DNA repair and was recently identified as a vulnerability of cancer cells with MSI [27,50–52]. Cells with MSI are vulnerable to massive apoptosis in the absence of WRN function because of accumulation of long TA dinucleotide repeats that form secondary structures that stall DNA forks during replication [53]. Consistent with this mechanism, MSS cell lines are not dependent on WRN helicase function [52]. Indeed, the *BRAF* mutant colorectal cancer cell lines that show vulnerability to WRN knock-down are all MSI high, suggesting that this is the underlying molecular defect directly responsible, rather than *BRAF* mutations. However, given the frequent co-occurrence of the two alterations in cell lines and clinical colorectal cancers, pharmacologic inhibition of WRN helicase in these cancers can be envisioned and would be expected to spare normal cells without MSI.

The other recurrent preferentially essential gene discovered in *BRAF* mutated cell lines, CAD, possesses the three first enzymatic activities in the pathway of de novo pyrimidine nucleotide biosynthesis in a single polypeptide of 2225 amino acids [54]. CAD is regulated by phosphorylation by MAPK, which activates the enzyme to promote nucleotide synthesis [55]. This regulation makes CAD a target of the KRAS/BRAF/MAPK cascade in response to growth factor signaling and activates an enzymatic function that sustains nucleotide production required for cell proliferation. Moreover, in colorectal cancer, CAD is regulated by MYC and when the metabolic reprogramming observed in cancer cells as a result of MYC activation is inhibited, cell growth is blocked by shutting down CAD and other enzymes of pyrimidine biosynthesis [56]. In cancer cells with deregulated proliferation secondary to *BRAF* mutations, loss of CAD function would deprive them from the required de novo pyrimidine nucleotides with potential catastrophic consequences due to loss of the coordinated response to the metabolic needs derived by high cancer cell proliferation. Thus, pharmacologic CAD inhibition with novel inhibitors in development may represent a therapeutic target in *BRAF* mutated cells with concomitant *PIK3CA* mutations, given that MAPK signaling and MYC are regulated by the two oncogenes [57].

A final interesting finding of the current investigation with potential future therapeutic implications is the identification of a NUAK family kinase (NUAK) inhibitor as one of the top hits in the pan-cancer BRAF mutant cell line screening. NUAK1 and NUAK2 are AMPK (AMP-activated Protein Kinase) related kinases with diverse functions in cancer cells [58]. NUAK1 promotes motility, invasion, and metastases of cancer cells [59,60]. NUAK1 shows higher expression in advanced stage colorectal cancers and in biopsies from liver metastatic sites, compared to primary tumors [61]. An important role of the kinase has been described in cancer cells with oncogene MYC overexpression, related to protection from oxidative stress resulting from MYC activity [62]. Mechanistically, NUAK1 contributes to mitochondrial plasticity and adaptation which is critical for cells bearing induction of oxidative respiratory chain component proteins effectuated by MYC [63]. Only 2 colorectal cancer cell lines with *BRAF* mutations RKO and HT-29 show MYC amplifications and both are more sensitive to the NUAK inhibitor WZ4003 than the mean sensitivity of the *BRAF* mutant group of colorectal cancer cell lines. Although these observations are based on a small number of cell lines, they suggest that *BRAF* mutant colorectal cancers with concomitant aberrations increasing oxidative stress could be candidates for combination therapies with NUAK kinases inhibitors.

A limitation of the current study is that relies exclusively in in silico publicly available data and no further experimental confirmation was performed. In addition, in the drug sensitivity analysis based on GDSC, cell lines are exposed to the assayed drugs as monotherapies and no data exist to inform combination therapies. Combinations of targeted anti-neoplastic drug therapies are increasingly recognized as being necessary for improvement of response in cancers which accumulate molecular alterations over time for their survival. Another limitation of the current study is that the cell line data do not definitely allow differentiation of a direct dependency on *BRAF* or *PIK3CA* mutations versus indirect effects related to other vulnerabilities such as MSI commonly co-occurring in these cell lines as the example of WRN helicase dependency illustrates. Moreover, it is expected that additional vulnerabilities that are not revealed with the approach used here exist in *BRAF* mutant colorectal cancers. For example, RANBP2, a binding protein of RAN (RAS related nuclear protein), a small GTPase of the RAS family, has been proposed as essential for survival of *BRAF* V600E mutant colorectal cancer cells and cells with a similar genomic signature [64].

In conclusion, targeted therapies of colorectal cancers that possess *BRAF* mutations with or without *PIK3CA* mutations could be developed based on the global molecular environment of these cancers and based on vulnerabilities uncovered in in vitro models. It is reassuring for the validity of the vulnerabilities discovered from cell lines models, that some of them, such as, for example, the synthetic lethality of MSI and WRN helicase, had previously been reported in pertinent systems. Leads discussed here need to be confirmed in in vivo studies followed by human trials in the population of interest.

Funding: This research received no external funding.

Institutional Review Board Statement: This research does not involve human subjects or animals and IRB approval was not required or obtained.

Informed Consent Statement: Not applicable.

Data Availability Statement: There are no data available beyond data included in the manuscript.

Conflicts of Interest: The author declares no conflict of interest.

References

1. Siegel, R.L.; Miller, K.D.; Fuchs, H.E.; Jemal, A. Cancer statistics, 2022. *CA Cancer J. Clin.* **2022**, *72*, 7–33. [CrossRef] [PubMed]
2. Siegel, R.L.; Miller, K.D.; Goding Sauer, A.; Fedewa, S.A.; Butterly, L.F.; Anderson, J.C.; Cercek, A.; Smith, R.A.; Jemal, A. Colorectal cancer statistics, 2020. *CA Cancer J. Clin.* **2020**, *70*, 145–164. [CrossRef]
3. Riedesser, J.E.; Ebert, M.P.; Betge, J. Precision medicine for metastatic colorectal cancer in clinical practice. *Ther. Adv. Med. Oncol.* **2022**, *14*, 17588359211072703. [CrossRef] [PubMed]
4. Rankin, A.; Klempner, S.J.; Erlich, R.; Sun, J.X.; Grothey, A.; Fakih, M.; George, T.J., Jr.; Lee, J.; Ross, J.S.; Stephens, P.J.; et al. Broad Detection of Alterations Predicted to Confer Lack of Benefit from EGFR Antibodies or Sensitivity to Targeted Therapy in Advanced Colorectal Cancer. *Oncologist* **2016**, *21*, 1306–1314. [CrossRef]
5. Tabernero, J.; Grothey, A.; Van Cutsem, E.; Yaeger, R.; Wasan, H.; Yoshino, T.; Desai, J.; Ciardiello, F.; Loupakis, F.; Hong, Y.S.; et al. Encorafenib Plus Cetuximab as a New Standard of Care for Previously Treated *BRAF* V600E-Mutant Metastatic Colorectal Cancer: Updated Survival Results and Subgroup Analyses from the BEACON Study. *J. Clin. Oncol.* **2021**, *39*, 273–284. [CrossRef] [PubMed]
6. Siena, S.; Di Bartolomeo, M.; Raghav, K.; Masuishi, T.; Loupakis, F.; Kawakami, H.; Yamaguchi, K.; Nishina, T.; Fakih, M.; Elez, E.; et al. Trastuzumab deruxtecan (DS-8201) in patients with HER2-expressing metastatic colorectal cancer (DESTINY-CRC01): A multicentre, open-label, phase 2 trial. *Lancet Oncol.* **2021**, *22*, 779–789. [CrossRef]
7. André, T.; Shiu, K.K.; Kim, T.W.; Jensen, B.V.; Jensen, L.H.; Punt, C.; Smith, D.; Garcia-Carbonero, R.; Benavides, M.; Gibbs, P.; et al. KEYNOTE-177 Investigators. Pembrolizumab in Microsatellite-Instability-High Advanced Colorectal Cancer. *N. Engl. J. Med.* **2020**, *383*, 2207–2218. [CrossRef] [PubMed]
8. Ratti, M.; Grizzi, G.; Passalacqua, R.; Lampis, A.; Cereatti, F.; Grassia, R.; Hahne, J.C. NTRK fusions in colorectal cancer: Clinical meaning and future perspective. *Expert Opin. Ther. Targets* **2021**, *25*, 677–683. [CrossRef] [PubMed]
9. Ji, J.; Wang, C.; Fakih, M. Targeting *KRAS*^{G12C}-Mutated Advanced Colorectal Cancer: Research and Clinical Developments. *Onco. Targets Ther.* **2022**, *15*, 747–756. [CrossRef]
10. Rosati, G.; Aprile, G.; Colombo, A.; Cordio, S.; Giampaglia, M.; Cappetta, A.; Porretto, C.M.; De Stefano, A.; Bilancia, D.; Avallone, A. Colorectal Cancer Heterogeneity and the Impact on Precision Medicine and Therapy Efficacy. *Biomedicines* **2022**, *10*, 1035. [CrossRef]
11. Ochsenreither, S.; Fiedler, W.M.; Conte, G.D.; Macchini, M.; Matos, I.; Habel, B.; Ahrens-Fath, I.; Raspagliesi, F.; Lorusso, D.; Keilholz, U.; et al. Safety and preliminary activity results of the GATTO study, a phase Ib study combining the anti-TA-MUC1 antibody gatipotuzumab with the anti-EGFR tomuzotuximab in patients with refractory solid tumors. *ESMO Open* **2022**, *7*, 100447. [CrossRef] [PubMed]
12. Grothey, A.; Fakih, M.; Tabernero, J. Management of BRAF-mutant metastatic colorectal cancer: A review of treatment options and evidence-based guidelines. *Ann. Oncol.* **2021**, *32*, 959–967. [CrossRef]
13. Kayhanian, H.; Goode, E.; Sclafani, F.; Ang, J.E.; Gerlinger, M.; Gonzalez de Castro, D.; Shepherd, S.; Peckitt, C.; Rao, S.; Watkins, D.; et al. Treatment and Survival Outcome of BRAF-Mutated Metastatic Colorectal Cancer: A Retrospective Matched Case-Control Study. *Clin. Color. Cancer* **2018**, *17*, e69. [CrossRef] [PubMed]
14. Jones, J.C.; Renfro, L.A.; Al-Shamsi, H.O.; Schrock, A.B.; Rankin, A.; Zhang, B.Y.; Kasi, P.M.; Voss, J.S.; Leal, A.D.; Sun, J.; et al. $^{\text{Non-V600}}$ BRAF Mutations Define a Clinically Distinct Molecular Subtype of Metastatic Colorectal Cancer. *J. Clin. Oncol.* **2017**, *35*, 2624–2630. [CrossRef] [PubMed]
15. Sahin, I.H.; Klostergaard, J. *BRAF* Mutations as Actionable Targets: A Paradigm Shift in the Management of Colorectal Cancer and Novel Avenues. *JCO Oncol. Pract.* **2021**, *17*, 723–730. [CrossRef] [PubMed]

16. Wan, P.T.; Garnett, M.J.; Roe, S.M.; Lee, S.; Niculescu-Duvaz, D.; Good, V.M.; Jones, C.M.; Marshall, C.J.; Springer, C.J.; Barford, D.; et al. Mechanism of activation of the RAF-ERK signaling pathway by oncogenic mutations of B-RAF. *Cell* **2004**, *116*, 855–867. [CrossRef]
17. Cancer Genome Atlas Network. Comprehensive molecular characterization of human colon and rectal cancer. *Nature* **2012**, *487*, 330–337. [CrossRef] [PubMed]
18. Giannakis, M.; Mu, X.J.; Shukla, S.A.; Qian, Z.R.; Cohen, O.; Nishihara, R.; Bahl, S.; Cao, Y.; Amin-Mansour, A.; Yamauchi, M.; et al. Genomic Correlates of Immune-Cell Infiltrates in Colorectal Carcinoma. *Cell Rep.* **2016**, *15*, 857–865. [CrossRef] [PubMed]
19. Yaeger, R.; Chatila, W.K.; Lipsyc, M.D.; Hechtman, J.F.; Cercek, A.; Sanchez-Vega, F.; Jayakumaran, G.; Middha, S.; Zehir, A.; Donoghue, M.T.A.; et al. Clinical Sequencing Defines the Genomic Landscape of Metastatic Colorectal Cancer. *Cancer Cell.* **2018**, *33*, 125–136.e3. [CrossRef] [PubMed]
20. Voutsadakis, I.A. The Landscape of PIK3CA Mutations in Colorectal Cancer. *Clin. Color. Cancer* **2021**, *20*, 201–215. [CrossRef] [PubMed]
21. Barretina, J.; Caponigro, G.; Stransky, N.; Venkatesan, K.; Margolin, A.A.; Kim, S.; Wilson, C.J.; Lehár, J.; Kryukov, G.V.; Sonkin, D.; et al. The Cancer Cell Line Encyclopedia enables predictive modelling of anticancer drug sensitivity. *Nature* **2012**, *483*, 603–607. [CrossRef] [PubMed]
22. Gao, J.; Aksoy, B.A.; Dogrusoz, U.; Dresdner, G.; Gross, B.; Sumer, S.O.; Sun, Y.; Jacobsen, A.; Sinha, R.; Larsson, E.; et al. Integrative analysis of complex cancer genomics and clinical profiles using the cBioPortal. *Sci. Signal.* **2013**, *6*, pl1. [CrossRef]
23. Mermel, C.H.; Schumacher, S.E.; Hill, B.; Meyerson, M.L.; Beroukhim, R.; Getz, G. GISTIC2.0 facilitates sensitive and confident localization of the targets of focal somatic copy-number alteration in human cancers. *Genome Biol.* **2011**, *12*, R41. [CrossRef]
24. Li, B.; Dewey, C.N. RSEM: Accurate transcript quantification from RNA-Seq data with or without a reference genome. *BMC Bioinform.* **2011**, *12*, 323. [CrossRef] [PubMed]
25. Chakravarty, D.; Gao, J.; Phillips, S.M.; Kundra, R.; Zhang, H.; Wang, J.; Rudolph, J.E.; Yaeger, R.; Soumerai, T.; Nissan, M.H.; et al. OncoKB: A Precision Oncology Knowledge Base. *JCO Precis. Oncol.* **2017**, *2017*, PO.17.00011. [CrossRef] [PubMed]
26. Iorio, F.; Knijnenburg, T.A.; Vis, D.J.; Bignell, G.R.; Menden, M.P.; Schubert, M.; Aben, N.; Gonçalves, E.; Barthorpe, S.; Lightfoot, H.; et al. A Landscape of Pharmacogenomic Interactions in Cancer. *Cell* **2016**, *166*, 740–754. [CrossRef]
27. Behan, F.M.; Iorio, F.; Picco, G.; Gonçalves, E.; Beaver, C.M.; Migliardi, G.; Santos, R.; Rao, Y.; Sassi, F.; Pinnelli, M.; et al. Prioritization of cancer therapeutic targets using CRISPR-Cas9 screens. *Nature* **2019**, *568*, 511–516. [CrossRef]
28. Boehm, J.S.; Garnett, M.J.; Adams, D.J.; Francies, H.E.; Golub, T.R.; Hahn, W.C.; Iorio, F.; McFarland, J.M.; Parts, L.; Vazquez, F. Cancer research needs a better map. *Nature* **2021**, *589*, 514–516. [CrossRef]
29. van der Meer, D.; Barthorpe, S.; Yang, W.; Lightfoot, H.; Hall, C.; Gilbert, J.; Francies, H.E.; Garnett, M.J. Cell Model Passports-a hub for clinical, genetic and functional datasets of preclinical cancer models. *Nucleic Acids Res.* **2019**, *47*, D923–D929. [CrossRef]
30. Tsherniak, A.; Vazquez, F.; Montgomery, P.G.; Weir, B.A.; Kryukov, G.; Cowley, G.S.; Gill, S.; Harrington, W.F.; Pantel, S.; Krill-Burger, J.M.; et al. Defining a Cancer Dependency Map. *Cell* **2017**, *170*, 564–576.e16. [CrossRef]
31. Marcotte, R.; Sayad, A.; Brown, K.R.; Sanchez-Garcia, F.; Reimand, J.; Haider, M.; Virtanen, C.; Bradner, J.E.; Bader, G.D.; Mills, G.B.; et al. Functional Genomic Landscape of Human Breast Cancer Drivers, Vulnerabilities, and Resistance. *Cell* **2016**, *164*, 293–309. [CrossRef] [PubMed]
32. Dwane, L.; Behan, F.M.; Gonçalves, E.; Lightfoot, H.; Yang, W.; van der Meer, D.; Shepherd, R.; Pignatelli, M.; Iorio, F.; Garnett, M.J. Project Score database: A resource for investigating cancer cell dependencies and prioritizing therapeutic targets. *Nucleic Acids Res.* **2021**, *49*, D1365–D1372. [CrossRef] [PubMed]
33. Meyers, R.M.; Bryan, J.G.; McFarland, J.M.; Weir, B.A.; Sizemore, A.E.; Xu, H.; Dharia, N.V.; Montgomery, P.G.; Cowley, G.S.; Pantel, S.; et al. Computational correction of copy number effect improves specificity of CRISPR-Cas9 essentiality screens in cancer cells. *Nat. Genet.* **2017**, *49*, 1779–1784. [CrossRef]
34. Dempster, J.M.; Boyle, I.; Vazquez, F.; Root, D.E.; Boehm, J.S.; Hahn, W.C.; Tsherniak, A.; McFarland, J.M. Chronos: A cell population dynamics model of CRISPR experiments that improves inference of gene fitness effects. *Genome Biol.* **2021**, *22*, 343. [CrossRef] [PubMed]
35. Poulikakos, P.I.; Sullivan, R.J.; Yaeger, R. Molecular Pathways and Mechanisms of BRAF in Cancer Therapy. *Clin Cancer Res.* **2022**, OF1–OF11. [CrossRef]
36. Zehir, A.; Benayed, R.; Shah, R.H.; Syed, A.; Middha, S.; Kim, H.R.; Srinivasan, P.; Gao, J.; Chakravarty, D.; Devlin, S.M.; et al. Mutational landscape of metastatic cancer revealed from prospective clinical sequencing of 10,000 patients. *Nat. Med.* **2017**, *23*, 703–713. [CrossRef]
37. Fruman, D.A.; Chiu, H.; Hopkins, B.D.; Bagrodia, S.; Cantley, L.C.; Abraham, R.T. The PI3K Pathway in Human Disease. *Cell* **2017**, *170*, 605–635. [CrossRef] [PubMed]
38. Solit, D.B.; Garraway, L.A.; Pratilas, C.A.; Sawai, A.; Getz, G.; Basso, A.; Ye, Q.; Lobo, J.M.; She, Y.; Osman, I.; et al. BRAF mutation predicts sensitivity to MEK inhibition. *Nature* **2006**, *439*, 358–362. [CrossRef]
39. Kopetz, S.; Grothey, A.; Yaeger, R.; Van Cutsem, E.; Desai, J.; Yoshino, T.; Wasan, H.; Ciardiello, F.; Loupakis, F.; Hong, Y.S.; et al. Encorafenib, Binimetinib, and Cetuximab in *BRAF* V600E-Mutated Colorectal Cancer. *N. Engl. J. Med.* **2019**, *381*, 1632–1643. [CrossRef]

40. van Geel, R.M.J.M.; Tabernero, J.; Elez, E.; Bendell, J.C.; Spreafico, A.; Schuler, M.; Yoshino, T.; Delord, J.P.; Yamada, Y.; Lolkema, M.P.; et al. A Phase Ib Dose-Escalation Study of Encorafenib and Cetuximab with or without Alpelisib in Metastatic *BRAF*-Mutant Colorectal Cancer. *Cancer Discov.* **2017**, *7*, 610–619. [CrossRef] [PubMed]
41. Voutsadakis, I.A. The genomic environment of *BRAF* mutated and *BRAF/PIK3CA* double mutated colorectal cancers. *J. Clin. Med.* **2022**, *11*, 5132. [CrossRef] [PubMed]
42. Tong, K.; Kothari, O.A.; Haro, K.S.; Panda, A.; Bandari, M.M.; Carrick, J.N.; Hur, J.J.; Zhang, L.; Chan, C.S.; Xing, J.; et al. SMAD4 is critical in suppression of BRAF-V600E serrated tumorigenesis. *Oncogene* **2021**, *40*, 6034–6048. [CrossRef] [PubMed]
43. Spender, L.C.; Ferguson, G.J.; Liu, S.; Cui, C.; Girotti, M.R.; Sibbet, G.; Higgs, E.B.; Shuttleworth, M.K.; Hamilton, T.; Lorigan, P.; et al. Mutational activation of BRAF confers sensitivity to transforming growth factor beta inhibitors in human cancer cells. *Oncotarget* **2016**, *7*, 81995–82012. [CrossRef]
44. Chen, A.J.; Zhou, G.; Juan, T.; Colicos, S.M.; Cannon, J.P.; Cabriera-Hansen, M.; Meyer, C.F.; Jurecic, R.; Copeland, N.G.; Gilbert, D.J.; et al. The dual specificity JKAP specifically activates the c-Jun N-terminal kinase pathway. *J. Biol. Chem.* **2002**, *277*, 36592–36601. [CrossRef] [PubMed]
45. Yu, D.; Li, Z.; Gan, M.; Zhang, H.; Yin, X.; Tang, S.; Wan, L.; Tian, Y.; Zhang, S.; Zhu, Y.; et al. Decreased expression of dual specificity phosphatase 22 in colorectal cancer and its potential prognostic relevance for stage IV CRC patients. *Tumour Biol.* **2015**, *36*, 8531–8535. [CrossRef] [PubMed]
46. Bradshaw, T.Y.; Romano, L.E.; Duncan, E.J.; Nethisinghe, S.; Abeti, R.; Michael, G.J.; Giunti, P.; Vermeer, S.; Chapple, J.P. A reduction in Drp1-mediated fission compromises mitochondrial health in autosomal recessive spastic ataxia of Charlevoix Saguenay. *Hum. Mol. Genet.* **2016**, *25*, 3232–3244. [CrossRef]
47. Voutsadakis, I.A. Polymerase epsilon mutations and concomitant β2-microglobulin mutations in cancer. *Gene* **2018**, *647*, 31–38. [CrossRef] [PubMed]
48. Riemer, P.; Sreekumar, A.; Reinke, S.; Rad, R.; Schäfer, R.; Sers, C.; Bläker, H.; Herrmann, B.G.; Morkel, M. Transgenic expression of oncogenic BRAF induces loss of stem cells in the mouse intestine, which is antagonized by β-catenin activity. *Oncogene* **2015**, *34*, 3164–3175. [CrossRef]
49. Morkel, M.; Riemer, P. Cell hierarchies in colorectal cancer: Focus on APC and BRAF. *Oncoscience* **2015**, *2*, 675–676. [CrossRef] [PubMed]
50. Chan, E.M.; Shibue, T.; McFarland, J.M.; Gaeta, B.; Ghandi, M.; Dumont, N.; Gonzalez, A.; McPartlan, J.S.; Li, T.; Zhang, Y.; et al. WRN helicase is a synthetic lethal target in microsatellite unstable cancers. *Nature* **2019**, *568*, 551–556. [CrossRef]
51. Lieb, S.; Blaha-Ostermann, S.; Kamper, E.; Rippka, J.; Schwarz, C.; Ehrenhöfer-Wölfer, K.; Schlattl, A.; Wernitznig, A.; Lipp, J.J.; Nagasaka, K.; et al. Werner syndrome helicase is a selective vulnerability of microsatellite instability-high tumor cells. *eLife* **2019**, *8*, e43333. [CrossRef] [PubMed]
52. Kategaya, L.; Perumal, S.K.; Hager, J.H.; Belmont, L.D. Werner Syndrome Helicase Is Required for the Survival of Cancer Cells with Microsatellite Instability. *iScience* **2019**, *13*, 488–497. [CrossRef]
53. van Wietmarschen, N.; Sridharan, S.; Nathan, W.J.; Tubbs, A.; Chan, E.M.; Callen, E.; Wu, W.; Belinky, F.; Tripathi, V.; Wong, N.; et al. Repeat expansions confer WRN dependence in microsatellite-unstable cancers. *Nature* **2020**, *586*, 292–298. [CrossRef]
54. Del Caño-Ochoa, F.; Ramón-Maiques, S. Deciphering CAD: Structure and function of a mega-enzymatic pyrimidine factory in health and disease. *Protein Sci.* **2021**, *30*, 1995–2008. [CrossRef] [PubMed]
55. Graves, L.M.; Guy, H.I.; Kozlowski, P.; Huang, M.; Lazarowski, E.; Pope, R.M.; Collins, M.A.; Dahlstrand, E.N.; Earp, H.S., 3rd; Evans, D.R. Regulation of carbamoyl phosphate synthetase by MAP kinase. *Nature* **2000**, *403*, 328–332. [CrossRef] [PubMed]
56. Satoh, K.; Yachida, S.; Sugimoto, M.; Oshima, M.; Nakagawa, T.; Akamoto, S.; Tabata, S.; Saitoh, K.; Kato, K.; Sato, S.; et al. Global metabolic reprogramming of colorectal cancer occurs at adenoma stage and is induced by MYC. *Proc. Natl. Acad. Sci. USA* **2017**, *114*, E7697–E7706. [CrossRef]
57. Lei, Z.; Wang, B.; Lu, Z.; Wang, N.; Tan, H.; Zheng, J.; Jia, Z. New regulatory mechanism-based inhibitors of aspartate transcarbamoylase for potential anticancer drug development. *FEBS J.* **2020**, *287*, 3579–3599. [CrossRef]
58. Port, J.; Muthalagu, N.; Raja, M.; Ceteci, F.; Monteverde, T.; Kruspig, B.; Hedley, A.; Kalna, G.; Lilla, S.; Neilson, L.; et al. Colorectal Tumors Require NUAK1 for Protection from Oxidative Stress. *Cancer Discov.* **2018**, *8*, 632–647. [CrossRef] [PubMed]
59. Kusakai, G.; Suzuki, A.; Ogura, T.; Kaminishi, M.; Esumi, H. Strong association of ARK5 with tumor invasion and metastasis. *J. Exp. Clin. Cancer Res.* **2004**, *23*, 263–268.
60. Zagórska, A.; Deak, M.; Campbell, D.G.; Banerjee, S.; Hirano, M.; Aizawa, S.; Prescott, A.R.; Alessi, D.R. New roles for the LKB1-NUAK pathway in controlling myosin phosphatase complexes and cell adhesion. *Sci. Signal.* **2010**, *3*, ra25. [CrossRef]
61. Kusakai, G.; Suzuki, A.; Ogura, T.; Miyamoto, S.; Ochiai, A.; Kaminishi, M.; Esumi, H. ARK5 expression in colorectal cancer and its implications for tumor progression. *Am. J. Pathol.* **2004**, *164*, 987–995. [CrossRef]
62. Liu, L.; Ulbrich, J.; Müller, J.; Wüstefeld, T.; Aeberhard, L.; Kress, T.R.; Muthalagu, N.; Rycak, L.; Rudalska, R.; Moll, R.; et al. Deregulated MYC expression induces dependence upon AMPK-related kinase 5. *Nature* **2012**, *483*, 608–612. [CrossRef] [PubMed]
63. Li, F.; Wang, Y.; Zeller, K.I.; Potter, J.J.; Wonsey, D.R.; O'Donnell, K.A.; Kim, J.W.; Yustein, J.T.; Lee, L.A.; Dang, C.V. Myc stimulates nuclearly encoded mitochondrial genes and mitochondrial biogenesis. *Mol. Cell. Biol.* **2005**, *25*, 6225–6234. [CrossRef] [PubMed]
64. Vecchione, L.; Gambino, V.; Raaijmakers, J.; Schlicker, A.; Fumagalli, A.; Russo, M.; Villanueva, A.; Beerling, E.; Bartolini, A.; Mollevi, D.G.; et al. A Vulnerability of a Subset of Colon Cancers with Potential Clinical Utility. *Cell* **2016**, *165*, 317–330. [CrossRef] [PubMed]

medicina

MDPI

Review

Global Impact of COVID-19 on Colorectal Cancer Screening: Current Insights and Future Directions

Jonathan Kopel [1,*], Bojana Ristic [2], Gregory L. Brower [3] and Hemant Goyal [4]

1 Department of Medicine, Texas Tech University Health Sciences Center, Lubbock, TX 79430, USA
2 Cell Biology and Biochemistry, Texas Tech University Health Sciences Center, Lubbock, TX 79430, USA; bojana.ristic@outlook.com
3 Department of Medication Education, Texas Tech University Health Sciences Center, Lubbock, TX 79430, USA; greg.brower@ttuhsc.edu
4 The Wright Center for Graduate Medical Education, Scranton, PA 18501, USA; goyalh@thewrightcenter.org
* Correspondence: jonathan.kopel@ttuhsc.edu

Abstract: The coronavirus disease 2019 (COVID-19) pandemic has brought significant challenges to many aspects of healthcare delivery since the first reported case in early December 2019. Once in the body, SARS-CoV-2 can spread to other digestive organs, such as the liver, because of the presence of ACE2 receptors. Colorectal cancer (CRC) remains the second-leading cause of death in the United States (US). Therefore, individuals are routinely screened using either endoscopic methods (i.e., flexible sigmoidoscopy and colonoscopy) or stool-based tests, as per the published guidelines. At the beginning of the COVID-19 pandemic, the Centers for Medicare and Medicaid Services (CMS) recommended that all non-urgent surgical and medical procedures, including screening colonoscopies, be delayed until the pandemic stabilization. This article aims to review the impact of COVID-19 on CRC screening.

Keywords: colorectal cancer screening; COVID-19; SARS-CoV-2

check for updates

Citation: Kopel, J.; Ristic, B.; Brower, G.L.; Goyal, H. Global Impact of COVID-19 on Colorectal Cancer Screening: Current Insights and Future Directions. *Medicina* **2022**, *58*, 100. https://doi.org/10.3390/medicina58010100

Academic Editors: Antonio M Scanu and Maria Rosaria De Miglio

Received: 18 December 2021
Accepted: 4 January 2022
Published: 10 January 2022

1. Introduction

The coronavirus disease 2019 (COVID-19) pandemic has brought significant challenges to many aspects of healthcare delivery since the first reported case in early December 2019 [1]. SARS-CoV-2 predominantly affects lungs, causing pneumonia, and acute respiratory distress syndrome (ARDS), but can also have extrapulmonary involvement, particularly with gastrointestinal (GI) symptoms [2]. GI symptoms, such as diarrhea (2–10.1%), nausea, and vomiting (1–3.6%), occur with modest frequency in COVID-19 compared to the fever and pulmonary symptoms in most patients [3–5]. Although the pathogenesis is still being investigated, current data suggests that the primary step for SARS-CoV-2 entry into the enterocytes occurs via the angiotensin-converting enzyme 2 (ACE2) protein [6–11].

Colorectal cancer (CRC) remains the second-leading cause of death in the United States (US) [12]. Therefore, individuals are routinely screened using either endoscopic methods (i.e., flexible sigmoidoscopy and colonoscopy) or stool-based tests, as per the published guidelines [13–18]. In the United States, the U.S. Preventive Services Task Force (USPSTF) recommends screening for colorectal cancer in all adults aged 50 to 75 years as, well as for adults aged 45 to 49 years [18]. In addition, the USPSTF recommends that clinicians selectively offer screening for colorectal cancer in adults aged 76 to 85 years. At the beginning of the COVID-19 pandemic, the Centers for Medicare and Medicaid Services (CMS) recommended that all non-urgent surgical and medical procedures, including screening colonoscopies, be delayed until the pandemic stabilization [19]. This was also done to reduce the potential risk of exposure to SARS-CoV-2, given the virus is present in fecal matter from COVID-19 patients [1,20,21]. In response, there was a 90% decrease in CRC screenings, resulting in a 32% decrease in new CRC diagnoses, as well as a 53%

decline in CRC-related surgical procedures by mid-April 2020 [19]. Moreover, by April 2021, the routine screening colonoscopy rate remained 50% lower than the pre-pandemic times [19]. Other forms of cancer screening, such as mammograms and pap tests, also decreased during the pandemic [22,23]. This article aims to review the impact of COVID-19 on CRC screening.

2. Methods

A literature search was performed using PubMed, Embase, SCOPUS, OVID, and Web of Science databases up to June 2021 to identify articles related to CRC screening and COVID-19. The search words used were "colorectal cancer", "screening", "COVID-19", and "SARS-CoV-2" alone and in combination. The results of the search are shown in Figure 1. The inclusion criteria for the studies included retrospective, longitudinal, and randomized control studies using different colorectal screening methods (guaiac-based fecal occult blood test, fecal immunochemical test (FIT), FIT-DNA test, colonoscopy, and sigmoidoscopy) during the COVID-19 pandemic. The exclusion criteria included any studies that were case reports, supplements, abstracts, commentaries, or had the wrong study focus for this review. A total of 331 studies were identified after the initial search. On initial review of the title and abstracts, 242 manuscripts were excluded because of irrelevance to our topic of interest. A further full manuscript review of the remaining articles was performed. A total of 20 articles were finally included in this review after removing 4 abstracts, 1 supplement, 23 commentaries, and 5 irrelevant studies.

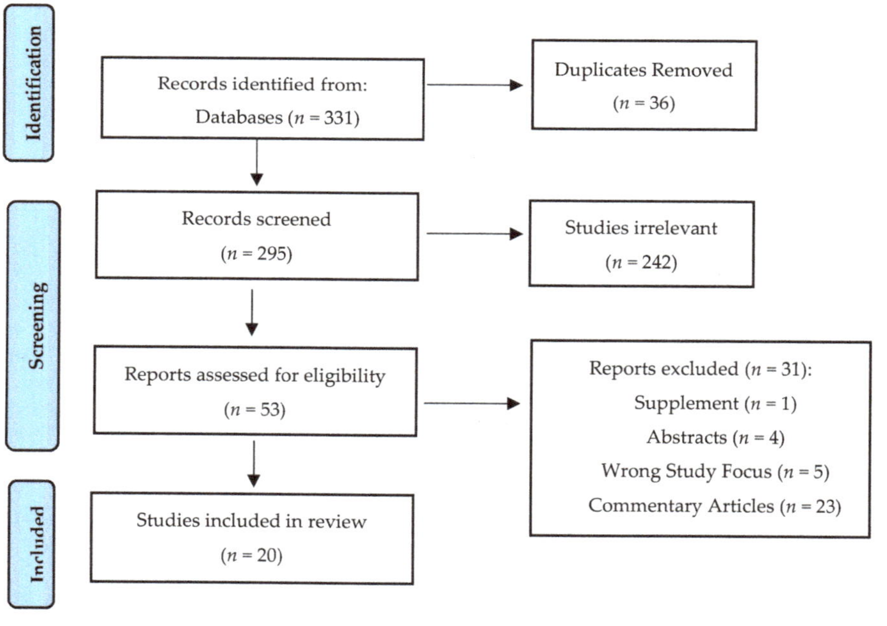

Figure 1. Systematic review article outline.

3. Effects of COVID-19 Pandemic on CRC Screening Pathway

The start of the COVID-19 pandemic disrupted several levels of primary care in the prevention of several preventable diseases, including CRC. Given the fear of transmitting SARS-CoV-2 in hospital settings, many elective procedures, such as colonoscopies, were discontinued until further notice. As a result, most primary care physicians utilized fecal immunochemical testing to continue providing CRC screening to prevent patients from traveling to the hospital and potentially exposing themselves to SARS-CoV-2. Amidst the

COVID-19 pandemic, the CRC screening program, which included initial fecal immunochemical testing (FIT) and diagnostic endoscopy, remained fully functional in the National Taiwan University Hospital in Northern Taiwan [24]. However, in comparison to the screening data collected from the corresponding quartal for previous years, the operations in this screening hub were interrupted [24]. This was observed due to the significant reduction in the number of patients that participated in the FIT screening, followed by the decrease in the immediate referrals to the diagnostic colonoscopy for the FIT-positive patients [24]. In addition, the already scheduled colonoscopy appointments had higher cancellation and rescheduling rates, with patients often listing the fear of nosocomial COVID-19 infection as the reason for not undergoing the procedure [24]. Furthermore, the number of diagnostic colonoscopies decreased drastically at the screening center in Japan during the state of emergency, which lasted 120 days [25]. The number of performed CRC surgeries, however, remained unchanged from the ones during previous years [25]. In comparison to the corresponding time period for three years prior to the COVID-19 pandemic, the population-based study in Hong Kong reported a 58.8% reduction in the number of lower colonoscopies performed from October 2019 to March 2020, which resulted in a 37% decline in the diagnosis of novel CRC cases [26].

The low to middle-income countries that recently started implementing nationwide CRC screening programs, such as Paraguay, Thailand, Iran, and Malaysia, continued to offer CRC screening, diagnostic, and treatment procedures during the COVID-19 pandemic [27]. They reported that administrating screening tests and diagnostic services for the screen-positive individuals were from 20–90% of the pre-COVID times, whereas the status of treatment services for cancer patients was better, with 65–90% ratings [27]. The authors discussed that the ability of these screening programs to resume to full capacity or beyond would be significantly difficult for the low to middle-income countries than it would be for the high-income countries [27]. Thus, the pandemic will have a lasting effect on these programs [27].

Most of the facilities that offered CRC services in England and Wales recorded a reduction in patient referral, resulting in an alteration in the CRC treatment plans that were brought up either by the delayed start of the treatment due to the fear of infection, lack of tissue diagnosis and radiological staging, and/or the limited resources [28]. The population-based study conducted in England revealed that the COVID-19 lockdown for months of April, May, and June 2020 reduced the CRC diagnostic rates, as the 2-week-wait referrals significantly dropped by 23%, whereas the number of performed colonoscopies decreased by 46% [29]. Furthermore, the lockdown elicited a 19% reduction in implementing the appropriate 31-day treatment plans [29]. Due to this, this study reported that over 3500 patients missed early CRC diagnosis, and did not undergo potentially lifesaving procedures during the COVID-19 lockdown [29]. The CRC screening rates remained low during the lockdown even, when that was performed in the "COVID-19 free" facility, which did not admit COVID-19 positive patients, and ensured ongoing COVID-19 testing for the facility's staff and patients [30]. The COVID-19 pandemic also affected molecular diagnostic metastatic CRC testing, which included performing quantitative PCR and next-generation sequencing (NGS) for *KRAS/NRAS* hot spot mutations in the largest molecular diagnostic centers for cancer patients and high-risk individuals in Serbia [31]. The number of performed analyses for the metastatic CRC during the state of emergency drastically decreased by 46%, followed by a 15% reduction in the number of GI tract cancer patients presented to the tumor board for formulating and implementing further treatment plans [31]. This trend did not recover even after the state of emergency was lifted [31].

The total deficit for CRC screening in the US during the COVID-19 pandemic compared to 2019 was calculated to be 3.8 million cases for both men and women [32]. The screening rates, when compared to 2019, decreased drastically by 79.3% in April of 2020 as a result of the lockdown, and they started to increase in June and July; the increase, however, did not re-calibrate to the same number as it was in 2019/2018, which showed that compensation for the missed diagnosis was not possible at the moment (13.1% lower than in previous years) [32]. The sharpest decline in CRC screening during lockdown was recorded in the Northeast geographical region of the USA, and amongst the population with a higher socioeconomic status [32]. In part, this was due to a 43% reduction in total referrals of primary care, with a 79% decline in urgent referrals, 64% reduction in routine referrals, and 40% reduction in "urgent suspicion of cancer" (USOC) patient referrals [33]. Moreover, during lockdown months in a healthcare system in Los Angeles, the colonoscopy rates declined to about only 12 per week, compared to the 223 per week pre-pandemic, whereas the FIT dropped to 61 per week, compared to 154 per week pre-pandemic [33]. After the lockdown was lifted, the number of colonoscopy appointments recovered to the number before the pandemic, whereas the noninvasive FIT and stool DNA tests recovered, and exceeded the pre-pandemic numbers [34]. Lastly, the interruption in developing screening programs in the CRC "hotspot" Appalachian Kentucky region elicited a significant backlog, and furthered the barriers that the program already had to face during its development [35].

4. The Aftermath of the Halted CRC Screening Due to COVID-19

The immediate halt in CRC screening and diagnostic modalities as a response to the COVID-19 pandemic has already been reflected in the sudden decrease in the diagnosis of the novel CRC cases [26,29,32]. Attributable to the interruption in timely diagnostic colonoscopy, the number of urgent admissions ascribed to obstructive CRC significantly increased during this time [25]. Moreover, the detected CRC cases were more severe in the lockdown group than in the previous years (47% vs. 25%), as the high-risk adenomas were often larger than 10 mm, contained villous compartment, high-grade dysplasia, and were serrated. However, the low-risk adenoma detection rate decreased (9% vs. 22%). Altogether, the lockdown group exhibited higher CRC detection (8% vs. 1%) [30].

The interruption of the diagnostic colonoscopies will also elicit long-term effects. Lui et al. predicted that 6.4% of CRC would have higher stage shifting, with an increase in stage IV carcinomas [26]. If the reduction in diagnostic procedures was not met, and if it further reduced to 20%, the stage shifting would increase by 7.2% [26]. The four country-specific CRC microsimulation models revealed that the delay in CRC diagnosis due to the diagnostic interruptions of three, six, and twelve months would result in a significant increase in the number of CRC incidence and CRC-related deaths during the period between 2020 and 2050 [36]. It was estimated that the relative increase of twelve-month disruptions would result in 0.4–0.9% additional CRC cases and 0.8–1.2% additional CRC-related deaths in the Netherlands, 1.2% additional CRC cases and 2% additional CRC-related deaths in Australia, and 0.6% additional CRC cases and 0.8% additional CRC-related deaths in Canada [36]. This devastating statistic can be minimized and mitigated if urgent catch-up screenings are provided to the screening and diagnostic facilities [36]. The delays also existed in performing the diagnostic colonoscopy for the two-week referral patients who were flagged urgent due to positive FIT results and severe symptoms [37]. The UK modeling study estimated that failure to provide timely diagnostics to these patients attributable to the delays of two, four, and six months in the follow-up diagnostic colonoscopies will significantly increase CRC-related deaths, and will result in a loss of life years [37].

A recent study by Santoro et al. studied the global impact of the COVID-19 pandemic on CRC screening, using a 35-item survey to assess the impact of COVID-19 on preoperative assessment, elective surgery, and postoperative management of colorectal cancer patients [38]. Respondents were sorted into two groups for comparison: (1) "delay" group: pandemic-affected colorectal cancer care; and (2) "no delay" group: unaffected colorectal cancer treatment. A total of 1051 respondents from 84 countries completed the survey [38]. There were no significant variations in demographics between the delay (n = 745, 70.9%) and no delay (n = 306, 29.1%) groups. In the delayed group, 48.9% of respondents reported a change in the initial surgical plan, and 26.3% reported a shift from elective to urgent operations [38]. Reductions in interdisciplinary team meetings, and the relocation of hospital and staff resources were significantly associated with delays in endoscopy, radiology, surgery, histopathology, and prolonged chemoradiation therapy-to-surgery intervals [38].

Furthermore, the status of the epidemic was linked to a patient's overall recovery during colorectal cancer treatment. Overall, there were noticeable improvements in colorectal cancer diagnostic and treatment procedures across the world. Rather than geographic variables, changes in CRC screening were linked to disparities in health care delivery systems, hospital preparation, resource availability, and local coronavirus illness 2019 prevalence [38].

A similar study conducted in the United Kingdom by the COVIDSurg Collaborative investigated the impact of SARS-CoV-2 on mortality after surgical resection of CRC during the early stages of the COVID-19 pandemic on surgical practice [39]. The COVIDSurg Collaborative used an international cohort study of patients who had colon or rectal cancer, and were not suspected of having SARS-CoV-2 before surgery. Using 2073 patients from 40 nations, the study found that 1.3 percent (27/2073) had a defunct stoma, and 3.0 percent (63/2073) had an end stoma rather than just an anastomosis. Thirty-day mortality was 1.8 percent (38/2073), with a 3.8 percent (78/2073) incidence of postoperative SARS-CoV-2, and a 4.9 percent (86/1738) anastomotic leak rate [39]. Patients without a leak or SARS-CoV-2 had the lowest mortality rate (14/1601, 0.9%), whereas patients with both a leak and SARS-CoV-2 had the greatest mortality rate (5/13, 38.5%) [39]. In contrast, anastomotic leak (adjusted odds ratio 6.01, 95 percent confidence interval 2.58–14.06), postoperative SARSCoV-2 (16.90, 7.86–36.38), male sex (2.46, 1.01–5.93), age > 70 years (2.87, 1.32–6.20), and advanced cancer stage (3.43, 1.16–10.21) were all independently linked with mortality [39]. There were fewer anastomotic leaks (4.9 percent versus 7.7%), and an average shorter duration of stay (6 versus 7 days) compared to pre-pandemic data, but increased mortality (1.7 percent versus 1.1 percent) [39]. Based on the patient, operational, and organizational risks, the COVIDSurg Collaborative suggested surgeons should take additional precautions against SARS-CoV-2 and anastomotic leak when performing surgery during the present and future COVID-19 waves.

5. Modified CRC Screening Approaches during COVID-19 Pandemic

In order to mitigate the survival decline attributed to the interruption of the CRC screening pathway elicited by the COVID-19 pandemic, public health screening programs need to be restructured [40]. Loveday et al. noted that prioritizing the high-risk CRC patients via FIT triage would alleviate 89% of deaths that would occur due to the lockdown backlog [37]. In turn, this strategy would reduce the nosocomial COVID-19 deaths [37]. Moreover, incorporating FIT screening as a triage tool for the 2-week wait referral patients elicited successful reallocation of the limited resources for high-risk CRC patients [41]. This mitigation approach was predicted to provide the additional CRC screening to approximately 588,800 novel patients, and establish about 2899 new CRC diagnoses, out of which 68.9% would be early-stage [19]. Furthermore, the utilization of other noninvasive stool-based DNA tests was marked effective in identifying high-risk patients, and prioritizing them for diagnostic colonoscopy and CRC treatment [34,35]. Miller et al. designed and implemented a novel CRC triage procedure named "COVID-adapted pathway", which successfully ameliorated the adverse effects of the diagnostic colonoscopy backlog [33]. According to this, the USOC patients referred by a general practitioner were triaged based on the severity of their symptoms [33]. In particular, patients with high-risk symptoms were triaged to CT with oral contrast and the quantitative FIT (qFIT), patients with low-risk symptoms were triaged with qFIT alone, whereas patients with palpable mass were outpatients [33]. Using this method, the number of detected cancers was similar to the previous year, while keeping the patients negative for COVID-19 [33]. The authors advised that the qFIT testing needed to be repeated twice, two weeks apart, because they noted occasional variations in results [33].

Furthermore, to ensure that the CRC screening continued, these programs needed to adopt new strategies, such as distributing stool-based CRC screening tests via mail, and shifting from paper to digital educational tools [27,35]. Furthermore, this included telehealth, such as teleconsultation of screen-positive individuals, conducting meetings over the phone, and dedicating call/text centers that would facilitate appointment scheduling [27,32,35]. Human factors, such as proper leadership and the allocation of resources, improved the overall outcome [28].

Lastly, the existence of "cold sites", such as CRC screening hubs, significantly improved the CRC screening and diagnosis pathway [28,30]. "Cold sites" were defined as the COVID-19 free clinics that were ensured via continuous SARS-CoV-2 testing, segregation of the emergency versus elective procedures, and the geographical separation of the COVID-19 facilities [28,30]. The continued screening in COVID-19 free clinics amidst the infectious agent pandemic was not only effective and necessary, but was also marked as safe, as there were no nosocomial infections related to COVID-19 in patients or in medical staff, as the safety procedures were followed closely [30].

6. Effects of COVID-19 Pandemic on Other Cancer Screenings

Since the referrals via the 2-week-wait urgent pathway dropped by 84% in the United Kingdom during the COVID-19 state of emergency, the delays of 3 months during the lockdown and the backlog of the referrals would decrease the 10-year survival by 10%, whereas the 6-month delays would decrease it to 30% for patients that suffer from carcinoma of the colon, rectum, esophagus, lung, liver, bladder, pancreas, stomach, larynx, and oropharynx [42]. Additional delays would further reduce the survival capacity while addressing the backlogs, and prioritizing the high-risk patients would mitigate these statistics [42]. Moreover, fewer breast cancer diagnoses were recorded in the Netherlands due to the suspension of the national screening programs [43]. In addition, low to middle-income countries, on average, reported ratings of less or equal to 50% of the pre-COVID-19 capacities by 61.1% of the participants for screening services, 44.4% of participants for diagnostic services, and 22.2% participants for treatment services for breast and cervical cancers [27]. To ensure the continuity of these national screening and diagnostic cancer programs, this study highlighted the necessity of incorporating the public community outreach through the expansion of the telehealth program [27]. Moreover, the COVID-19 pandemic interrupted the molecular diagnostic testing for non-small cell lung carcinoma (*EGRF* mutations) and metastatic melanoma (*BRAF* mutations), and hereditary breast/ovarian cancer (*BRCA1/2* mutations) in the largest molecular diagnostic center for cancer patients in Serbia drastically decreased [31]. The decline did not recover after the state of emergency was lifted [31]. Lastly, lockdown months exhibited a sharp reduction in screening rates for breast (90.8%) and prostate cancer (63.4%) [32]. It was estimated that about 3.9 million women were undiagnosed for breast cancer due to this interruption, and 1.6 million men missed diagnosis for prostate cancer in the US [32]. The highest decline was recorded for the northwest US region, and amongst the population with a high socioeconomic status [32].

7. Clinical Implications and Limitations

There is no doubting that COVID-19 has a significant impact on cancer screening. The CRC death rates are expected to rise as a result of screening test disruptions and delays. Although the outlook for CRC appears bleak, there are several lessons to be learned from the COVID-19 pandemic in terms of CRC screening, diagnosis, and overall prevention. More use of alternate techniques, such as FIT testing, are important for the future of CRC screening and diagnosis. Furthermore, the availability of FIT testing can help to reduce racial health inequalities. Despite the fact that routine screening techniques are beginning to resume as COVID-19 vaccinations are being provided in many industrialized countries, the pandemic has changed the way healthcare practitioners perceive CRC screening. Though colonoscopy will always be the gold standard for CRC screening, FIT tests and other screening procedures offer considerable strengths and unique qualities that make them useful and ideal in certain scenarios. It is critical that we use what we know about COVID-19's impact on CRC to plan for and prevent human suffering in future pandemics and other public health emergencies. Early detection, combined with adequate care, saves lives, and doing so might help avoid the impact of a future occurrence on CRC from being as disastrous. Regular screening techniques have resumed in recent years, since COVID-19 vaccinations have become more widely available. However, the previously described delays continue to pay a toll on human lives in terms of the delayed diagnosis and progression of CRC.

This study has several limitations. First, the long-term effects of COVID-19 are still being investigated for CRC surveillance and other cancers. The full impact of the pandemic on cancer screening will be understood in the future. Second, the studies did not adequately sample the effects of the COVID-19 pandemic on CRC screening across the globe. This is likely due to differences in severity and resource limitations in acquiring and maintaining data on CRC screening during the pandemic. Third, not all countries equally report CRC screening rates that can be compared between countries. Fourth, different studies reported different methods of CRC screening, which may influence the overall rate of CRC before, during, and after the pandemic. Further longitudinal studies will be needed to address the limitations mentioned in this paper.

8. Conclusions

Continuous CRC screening efforts, from population-wide stool-based testing to diagnostic endoscopies and treatments, have elicited early cancer detection, and improved the devastating statistics regarding the CRC diagnosis outcome. However, programs are not available everywhere, and, in some places, are not efficient. Therefore, they require constant improvements in terms of encouraging patient participation, educating the population, and stratifying low- and high-risk patients to ensure cancer diagnosis and prompt treatment. The COVID-19 pandemic caused the world to pause, and instituted lockdowns, notably interrupting CRC screening programs. The reasons for the halt were the allocation of limited hospital resources towards the fight against COVID-19, the ongoing fear of nosocomial SARS-CoV-2 infection, and the overall overwhelming burden that the pandemic placed onto the healthcare system. For CRC screening programs, this included a drop in referrals from a general practitioner, patients' unwillingness to partake in stool-based testing, canceling or rescheduling colonoscopy appointments by patients out of fear or by institutions because they worked in limited capacities, and changing treatment plans to comply with the pandemic-elicited regulations. Although some centers remained fully functional or adopted novel screening pathway procedures which included telehealth, the diagnostic capacities halted. In this manner, the substantial number of CRC patients went undiagnosed, which, in the short term, resulted in an increase of obstructive CRC, and the presence of high-risk adenomas. The long-term effects of the diagnosis backlog could result in a devastating rise of late-stage CRC cases, and the overall loss of life years due to the lack of appropriate treatments for these patients. These prognostics, however, can be mitigated if proper catch-up screenings are provided. These lessons can also serve as a teaching moment for healthcare leadership, and can provide guidelines for minimizing and altogether avoiding the interruption of cancer screening programs if novel pandemic-causing infectious agents appear.

Author Contributions: Conception and design: J.K. and H.G.; Collection and assembly of data: J.K.; Manuscript writing: J.K., B.R., G.L.B. and H.G.; Final approval of manuscript: J.K., B.R., G.L.B. and H.G. All authors have read and agreed to the published version of the manuscript.

Funding: This research received no external funding.

Institutional Review Board Statement: Not applicable.

Informed Consent Statement: Not applicable.

Data Availability Statement: Not applicable.

Conflicts of Interest: The authors declare no conflict of interest.

References

1. Kopel, J.; Perisetti, A.; Gajendran, M.; Boregowda, U.; Goyal, H. Clinical Insights into the Gastrointestinal Manifestations of COVID-19. *Am. J. Dig. Dis.* **2020**, *65*, 1932–1939. [CrossRef] [PubMed]
2. Wu, D.; Wu, T.; Liu, Q.; Yang, Z. The SARS-CoV-2 outbreak: What we know. *Int. J. Infect. Dis.* **2020**, *94*, 44–48. [CrossRef] [PubMed]
3. Wu, F.; Zhao, S.; Yu, B.; Chen, Y.-M.; Wang, W.; Song, Z.-G.; Hu, Y.; Tao, Z.-W.; Tian, J.-H.; Pei, Y.-Y.; et al. A new coronavirus associated with human respiratory disease in China. *Nature* **2020**, *579*, 265–269. [CrossRef] [PubMed]
4. Chen, N.; Zhou, M.; Dong, X.; Qu, J.; Gong, F.; Han, Y.; Qiu, Y.; Wang, J.; Liu, Y.; Wei, Y.; et al. Epidemiological and clinical characteristics of 99 cases of 2019 novel coronavirus pneumonia in Wuhan, China: A descriptive study. *Lancet* **2020**, *395*, 507–513. [CrossRef]
5. Ji, L.-N.; Chao, S.; Wang, Y.-J.; Li, X.-J.; Mu, X.-D.; Lin, M.-G.; Jiang, R.-M. Clinical features of pediatric patients with COVID-19: A report of two family cluster cases. *World J. Pediatr.* **2020**, *16*, 267–270. [CrossRef]
6. Zhang, J.-J.; Dong, X.; Cao, Y.Y.; Yuan, Y.D.; Yang, Y.B.; Yan, Y.Q.; Akdis, C.A.; Gao, Y.D. Clinical characteristics of 140 patients infected with SARS-CoV-2 in Wuhan, China. *Allergy* **2020**, *75*, 1730–1741. [CrossRef]
7. Joint GI Society Message: COVID-19 Clinical Insights for Our Community of Gastroenterologists and Gastroenterology Care Providers. Available online: https://gastro.org/press-releases/joint-gi-society-message-covid-19-clinical-insights-for-our-community-of-gastroenterologists-and-gastroenterology-care-providers/ (accessed on 3 January 2022).
8. Riordan, J.F. Angiotensin-I-converting enzyme and its relatives. *Genome Biol.* **2003**, *4*, 225. [CrossRef]
9. Xiao, F.; Tang, M.; Zheng, X.; Liu, Y.; Li, X.; Shan, H. Evidence for gastrointestinal infection of SARS-CoV-2. *Gastroenterology* **2020**, *158*, 1831–1833. [CrossRef]
10. Zhou, P.; Yang, X.-L.; Wang, X.-G.; Hu, B.; Zhang, L.; Zhang, W.; Si, H.-R.; Zhu, Y.; Li, B.; Huang, C.-L.; et al. A pneumonia outbreak associated with a new coronavirus of probable bat origin. *Nature* **2020**, *579*, 270–273. [CrossRef]
11. Hoffmann, M.; Kleine-Weber, H.; Schroeder, S.; Krüger, N.; Herrler, T.; Erichsen, S.; Schiergens, T.S.; Herrler, G.; Wu, N.; Nitsche, A.; et al. SARS-CoV-2 Cell Entry Depends on ACE2 and TMPRSS2 and Is Blocked by a Clinically Proven Protease Inhibitor. *Cell* **2020**, *181*, 271–280.e8. [CrossRef]
12. Siegel, R.L.; Miller, K.D.; Goding Sauer, A.; Jemal, A. Colorectal cancer statistics, 2020. *CA Cancer J. Clin.* **2020**, *70*, 145–164. [CrossRef]
13. Cyhaniuk, A.; Coombes, M.E. Longitudinal adherence to colorectal cancer screening guidelines. *Am. J. Manag. Care* **2016**, *22*, 105–111. [PubMed]
14. Baxter, N.N. Association of Colonoscopy and Death From Colorectal Cancer. *Ann. Intern. Med.* **2009**, *150*, 1–8. [CrossRef] [PubMed]
15. Heitman, S.J.; Hilsden, R.J.; Au, F.; Dowden, S.; Manns, B.J. Colorectal cancer screening for average-risk North Americans: An economic evaluation. *PLoS Med.* **2010**, *7*, e1000370. [CrossRef]
16. Mandel, J.S.; Bond, J.H.; Church, T.R.; Snover, D.C.; Bradley, G.M.; Schuman, L.M.; Ederer, F. Reducing Mortality from Colorectal Cancer by Screening for Fecal Occult Blood. *N. Engl. J. Med.* **1993**, *328*, 1365–1371. [CrossRef] [PubMed]
17. Shaukat, A.; Shyne, M.; Mandel, J.S.; Snover, D.; Church, T.R. Colonoscopy With Polypectomy Reduces Long-Term Incidence of Colorectal Cancer in Both Men and Women: Extended Results From the Minnesota Colon Cancer Control Study. *Gastroenterology* **2021**, *160*, 1397–1399.e3. [CrossRef] [PubMed]
18. Force, U.P.S.T. Screening for Colorectal Cancer: US Preventive Services Task Force Recommendation Statement. *JAMA* **2021**, *325*, 1965–1977.
19. Issaka, R.B.; Taylor, P.; Baxi, A.; Inadomi, J.M.; Ramsey, S.D.; Roth, J. Model-Based Estimation of Colorectal Cancer Screening and Outcomes During the COVID-19 Pandemic. *JAMA Netw. Open* **2021**, *4*, e216454. [CrossRef] [PubMed]
20. Mirjalali, H.; Nazemalhosseini-Mojarad, E.; Yadegar, A.; Mohebbi, S.R.; Baghaei, K.; Shahrokh, S.; Aghdaei, H.A.; Zali, M.R. The Necessity of Stool Examination in Asymptomatic Carriers as a Strategic Measure to Control Further Spread of SARS-CoV-2. *Front. Public Health* **2020**, *8*, 553589. [CrossRef]
21. Aloysius, M.M.; Thatti, A.; Gupta, A.; Sharma, N.; Bansal, P.; Goyal, H. COVID-19 presenting as acute pancreatitis. *Pancreatology* **2020**, *20*, 1026–1027. [CrossRef]
22. Dennis, L.K.; Hsu, C.-H.; Arrington, A.K. Reduction in Standard Cancer Screening in 2020 throughout the U.S. *Cancers* **2021**, *13*, 5918. [CrossRef] [PubMed]
23. Gallo, G.; La Torre, M.; Pietroletti, R.; Bianco, F.; Altomare, D.F.; Pucciarelli, S.; Gagliardi, G.; Perinotti, R. Italian society of colorectal surgery recommendations for good clinical practice in colorectal surgery during the novel coronavirus pandemic. *Tech. Coloproctol.* **2020**, *24*, 501–505. [CrossRef] [PubMed]
24. Cheng, S.; Chen, C.; He, H.; Chang, L.; Hsu, W.; Wu, M.; Chiu, H. Impact of COVID-19 pandemic on fecal immunochemical test screening uptake and compliance to diagnostic colonoscopy. *J. Gastroenterol. Hepatol.* **2020**, *36*, 1614–1619. [CrossRef] [PubMed]
25. Mizuno, R.; Ganeko, R.; Takeuchi, G.; Mimura, K.; Nakahara, H.; Hashimoto, K.; Hinami, J.; Shimomatsuya, T.; Kubota, Y. The number of obstructive colorectal cancers in Japan has increased during the COVID-19 pandemic: A retrospective single-center cohort study. *Ann. Med. Surg.* **2020**, *60*, 675–679. [CrossRef] [PubMed]

26. Lui, T.K.; Leung, K.; Guo, C.-G.; Tsui, V.W.; Wu, J.T.K.; Leung, W.K. Impacts of the Coronavirus 2019 Pandemic on Gastrointestinal Endoscopy Volume and Diagnosis of Gastric and Colorectal Cancers: A Population-Based Study. *Gastroenterology* **2020**, *159*, 1164–1166.e3. [CrossRef]
27. Villain, P.; Carvalho, A.L.; Lucas, E.; Mosquera, I.; Zhang, L.; Muwonge, R.; Selmouni, F.; Sauvaget, C.; Basu, P.; for the IARC COVID-19 Impact Study Group. Cross-sectional survey of the impact of the COVID-19 pandemic on cancer screening programs in selected low- and middle-income countries: Study from the IARC COVID-19 impact study group. *Int. J. Cancer* **2021**, *149*, 97–107. [CrossRef]
28. Boyle, J.M.; Kuryba, A.; Blake, H.A.; Aggarwal, A.; van der Meulen, J.; Walker, K.; Braun, M.; Fearnhead, N. The impact of the first peak of the COVID-19 pandemic on colorectal cancer services in England and Wales: A national survey. *Colorectal. Dis.* **2021**, *23*, 1733–1744. [CrossRef]
29. Morris, E.J.A.; Goldacre, R.; Spata, E.; Mafham, M.; Finan, P.J.; Shelton, J.; Richards, M.; Spencer, K.; Emberson, J.; Hollings, S.; et al. Impact of the COVID-19 pandemic on the detection and management of colorectal cancer in England: A population-based study. *Lancet Gastroenterol. Hepatol.* **2021**, *6*, 199–208. [CrossRef]
30. D'Ovidio, V.; Lucidi, C.; Bruno, G.; Lisi, D.; Miglioresi, L.; Bazuro, M.E. Impact of COVID-19 Pandemic on Colorectal Cancer Screening Program. *Clin. Colorectal. Cancer* **2021**, *20*, e5–e11. [CrossRef]
31. Cavic, M.; Krivokuca, A.; Boljevic, I.; Spasic, J.; Mihajlovic, M.; Pavlovic, M.; Damjanovic, A.; Radosavljevic, D.; Jankovic, R. Exploring the real-world effect of the SARS-CoV-2 pandemic on the molecular diagnostics for cancer patients and high-risk individuals. *Expert Rev. Mol. Diagn.* **2021**, *21*, 101–107. [CrossRef]
32. Chen, R.C.; Haynes, K.; Du, S.; Barron, J.; Katz, A.J. Association of Cancer Screening Deficit in the United States With the COVID-19 Pandemic. *JAMA Oncol.* **2021**, *7*, 878. [CrossRef] [PubMed]
33. Miller, J.; Maeda, Y.; Au, S.; Gunn, F.; Porteous, L.; Pattenden, R.; MacLean, P.; Noble, C.L.; Glancy, S.; Dunlop, M.G.; et al. Short-term outcomes of a COVID-adapted triage pathway for colorectal cancer detection. *Colorectal. Dis.* **2021**, *23*, 1639–1648. [CrossRef] [PubMed]
34. Myint, A.; Roh, L.; Yang, L.; Connolly, L.; Esrailian, E.; May, F.P. Noninvasive Colorectal Cancer Screening Tests Help Close Screening Gaps During Coronavirus Disease 2019 Pandemic. *Gastroenterology* **2021**, *161*, 712–714. [CrossRef] [PubMed]
35. Kruse-Diehr, A.J.; Dignan, M.; Cromo, M.; Carman, A.L.; Rogers, M.; Gross, D.; Russell, S. Building Cancer Prevention and Control Research Capacity in Rural Appalachian Kentucky Primary Care Clinics During COVID-19: Development and Adaptation of a Multilevel Colorectal Cancer Screening Project. *J. Cancer Educ.* **2021**, 1–7. [CrossRef] [PubMed]
36. De Jonge, L.; Worthington, J.; van Wifferen, F.; Iragorri, N.; Peterse, E.F.P.; Lew, J.; Greuter, M.J.E.; Smith, H.A.; Feletto, E.; Yong, J.H.E.; et al. Impact of the COVID-19 pandemic on faecal immunochemical test-based colorectal cancer screening programmes in Australia, Canada, and the Netherlands: A comparative modelling study. *Lancet Gastroenterol. Hepatol.* **2021**, *6*, 304–314. [CrossRef]
37. Loveday, C.; Sud, A.; Jones, M.; Broggio, J.; Scott, S.; Gronthound, F.; Torr, B.; Garrett, A.; Nicol, D.L.; Jhanji, S.; et al. Prioritisation by FIT to mitigate the impact of delays in the 2-week wait colorectal cancer referral pathway during the COVID-19 pandemic: A UK modelling study. *Gut* **2021**, *70*, 1053–1060. [CrossRef]
38. Santoro, G.A.; Grossi, U.; Murad-Regadas, S.; Nunoo-Mensah, J.W.; Mellgren, A.; Di Tanna, G.L.; Gallo, G.; Tsang, C.; Wexner, S.D. DElayed COloRectal cancer care during COVID-19 Pandemic (DECOR-19): Global perspective from an international survey. *Surgery* **2021**, *169*, 796–807. [CrossRef]
39. COVIDSurg Collaborative. Outcomes from elective colorectal cancer surgery during the SARS-CoV-2 pandemic. *Colorectal. Dis.* **2020**, *23*, 732–749.
40. Kadakuntla, A.; Wang, T.; Medgyesy, K.; Rrapi, E.; Litynski, J.; Adynski, G.; Tadros, M. Colorectal cancer screening in the COVID-19 era. *World J. Gastrointest. Oncol.* **2021**, *13*, 238–251. [CrossRef]
41. Maclean, W.; Limb, C.; Mackenzie, P.; Whyte, M.B.; Benton, S.C.; Rockall, T.; Jourdan, I. Adoption of faecal immunochemical testing for 2-week-wait colorectal patients during the COVID-19 pandemic: An observational cohort study reporting a new service at a regional centre. *Colorectal. Dis.* **2020**, *23*, 1622–1629. [CrossRef]
42. Sud, A.; Torr, B.; Jones, M.; Broggio, J.; Scott, S.; Loveday, C.; Garrett, A.; Gronthoud, F.; Nicol, D.L.; Jhanji, S.; et al. Effect of delays in the 2-week-wait cancer referral pathway during the COVID-19 pandemic on cancer survival in the UK: A modelling study. *Lancet Oncol.* **2020**, *21*, 1035–1044. [CrossRef]
43. Dinmohamed, A.G.; Cellamare, M.; Visser, O.; De Munck, L.; Elferink, M.A.G.; Westenend, P.J.; Wesseling, J.; Broeders, M.J.M.; Kuipers, E.J.; Merkx, M.A.W.; et al. The impact of the temporary suspension of national cancer screening programmes due to the COVID-19 epidemic on the diagnosis of breast and colorectal cancer in the Netherlands. *J. Hematol. Oncol.* **2020**, *13*, 147. [CrossRef] [PubMed]

medicina

MDPI

Article

BRAF, MEK, and EGFR Triplet Inhibitors as Salvage Therapy in *BRAF*-Mutated Metastatic Colorectal Cancer—A Case Series Study *Target Therapy of* BRAF-*Mutated mCRC*

Jen-Hao Yeh [1,2,3,4], Hsiang-Lin Tsai [5,6], Yen-Cheng Chen [5], Ching-Chun Li [5], Ching-Wen Huang [5,6], Tsung-Kun Chang [5], Wei-Chih Su [5], Po-Jung Chen [5], Yu-Peng Liu [1] and Jaw-Yuan Wang [1,5,6,7,8,9,*]

1 Graduate Institute of Clinical Medicine, College of Medicine, Kaohsiung Medical University, Kaohsiung 80708, Taiwan; b9202078@gmail.com (J.-H.Y.); ypliu@kmu.edu.tw (Y.-P.L.)
2 Division of Gastroenterology and Hepatology, Department of Internal Medicine, E-DA Dachang Hospital, Kaohsiung 80794, Taiwan
3 Department of Medical technology, College of Medicine, I-Shou University, Kaohsiung 82445, Taiwan
4 Division of Gastroenterology and Hepatology, Department of Internal Medicine, E-DA Hospital, Kaohsiung 82445, Taiwan
5 Division of Colorectal Surgery, Department of Surgery, Kaohsiung Medical University Hospital, Kaohsiung Medical University, Kaohsiung 80708, Taiwan; chunpin870132@yahoo.com.tw (H.-L.T.); googoogi05@gmail.com (Y.-C.C.); dobird05@yahoo.com.tw (C.-C.L.); baseball5824@yahoo.com.tw (C.-W.H.); tsungkunchang@gmail.com (T.-K.C.); lake0126@yahoo.com.tw (W.-C.S.); glaudiotennis@gmail.com (P.-J.C.)
6 Department of Surgery, Faculty of Medicine, College of Medicine, Kaohsiung Medical University, Kaohsiung 80708, Taiwan
7 Center for Cancer Research, Kaohsiung Medical University, Kaohsiung 80708, Taiwan
8 Center for Liquid biopsy and Cohort Research, Kaohsiung Medical University, Kaohsiung 80708, Taiwan
9 Pingtung Hospital, Ministry of Health and Welfare, Pingtung 90054, Taiwan
* Correspondence: cy614112@ms14.hinet.net; Tel.: +886-7-3122805; Fax: +886-7-3114679

Citation: Yeh, J.-H.; Tsai, H.-L.; Chen, Y.-C.; Li, C.-C.; Huang, C.-W.; Chang, T.-K.; Su, W.-C.; Chen, P.-J.; Liu, Y.-P.; Wang, J.-Y. BRAF, MEK, and EGFR Triplet Inhibitors as Salvage Therapy in *BRAF*-Mutated Metastatic Colorectal Cancer—A Case Series Study *Target Therapy of* BRAF-*Mutated mCRC*. *Medicina* **2021**, *57*, 1339. https://doi.org/10.3390/medicina57121339

Academic Editors: Antonio M. Scanu and Maria Rosaria De Miglio

Received: 8 November 2021
Accepted: 4 December 2021
Published: 7 December 2021

Abstract: *Background and objectives*: Patients with *BRAF*-mutated metastatic colorectal cancer have considerably poorer responses to conventional systemic treatment. The real-world effects of triplet therapy with BRAF, mitogen-activated protein kinase kinase, and epidermal growth factor receptor inhibitors in Asia have not been well-reported. *Materials and Methods*: This single-center case series included patients with *BRAF*-mutated metastatic colorectal cancer undergoing triplet therapy after failure of prior systemic treatment from 2016 to 2020. The primary outcome was progression-free survival, and secondary outcomes were overall survival, response rate, disease control rate, and adverse events. *Results*: Nine eligible patients with *BRAF*-mutated metastatic colorectal cancer receiving triplet therapy were enrolled, with a median follow-up time of 14.5 months (range, 1–26). Most patients (88.8%) had two or more prior systemic treatments, and the triplet regimen was mainly dabrafenib, trametinib, and panitumumab. The overall response rate and disease control rate were 11.1% and 33.3%, respectively. Median progression-free survival and overall survival were 2.9 and 7.4 months, respectively, and a trend toward better overall survival was found with left-sided metastatic colorectal cancer compared with right-sided disease (9.2 vs. 6.9 months, $p = 0.093$). Adverse events were mostly Grade 1–2, including nausea, hypertension, gastrointestinal symptoms, and skin disorders. *Conclusions*: In this single-center case series, triplet therapy with BRAF, mitogen-activated protein kinase kinase, and epidermal growth factor receptor inhibitors in *BRAF*-mutated metastatic colorectal cancer had an acceptable safety profile and reasonable efficacy.

Keywords: metastatic colorectal cancer; *BRAF* mutation; triple target therapy

1. Introduction

Cases of metastatic colorectal cancer (mCRC) comprise approximately one-fourth of all colorectal cancer (CRC) cases at initial diagnosis, and an additional 20% of CRC patients may also present subsequent metachronous metastasis despite treatment [1,2]. Progress has been made in various treatment strategies, including surgery, cytotoxic chemotherapy, target therapy, and immunotherapy. RAS wild type mCRC is still a treatment challenge, especially when other resistant gene alterations are present.

Along with RAS [3,4] and microsatellite instability [5,6], the BRAF V600E mutation [7,8] is a well-known biomarker that has an impact on mCRC survival and may affect the response of systemic and targeted therapies. Although the BRAF mutation is only detected in 5%–10% of all cases, mCRC patients who are microsatellite-stable with the BRAF V600E mutation have worse survival and response to anti-epidermal growth factor receptor (EGFR) agents [9,10]. However, resistance to anti-EGFR agents may be overcome with BRAF inhibitors [11,12], which may be beneficial in patients with progressive mCRC after the failure of first-line treatment.

The combination of a BRAF inhibitor and anti-EGFR agent, with and without a mitogen-activated protein kinase kinase (MEK) inhibitor, has been evaluated in several studies as a promising regimen for mCRC after first-line standard treatment [11,13,14]. The phase III BEACON trial demonstrated that the triplet regimen, which consists of a BRAF inhibitor, anti-EGFR agent, and MEK inhibitor, significantly improved overall survival (OS) and progression-free survival (PFS) compared with the control group (chemotherapy plus anti-EGFR agent) [11]. Another ongoing single-arm trial (ANCHOR CRC, a phase II study of first-line triple therapy with cetuximab, encorafenib, and binimetinib) also showed a favorable response rate [15]. However, real-world data on the triplet regimen as a later line of systemic treatment in Asian patients, is still lacking due to the scarcity of such patients. Thus, this case series aimed to report the clinical outcomes and safety of triplet therapy in mCRC patients with BRAF V600E mutations after the failure of at least first-line chemotherapy.

2. Methods

2.1. Patient Eligibility

This case series was a single-center study conducted at our hospital. Eligible cases were identified through medical chart review from April 2016 to April 2020. Patients were included if they met all the following criteria: (1) recurrence or progressive disease after first-line chemotherapy plus target therapy, with or without surgery; (2) at least one metastatic focus found in an imaging study; (3) pathologic examination of the tumor specimen revealing a BRAF V600E mutation; and (4) receiving triplet therapy as the second or later line of systemic treatment. Eligible cases were enrolled for this study until April 2020. This study was approved by the institutional review board of our hospital [KMUHIRB-2012-03-02(II)].

2.2. Analysis of BRAF Mutation, RAS Mutation, and Status of Microsatellite Stability

BRAF V600E mutation analysis was performed using direct deoxyribonucleic acid (DNA) sequencing from formalin-fixed, paraffin-embedded CRC tissue samples according to our previous study [16]. After deparaffinization and air-drying, DNA was isolated using the proteinase K and QIAamp DNA Micro Kit (QIAGEN). A high-resolution melting analysis was undertaken using the LightCycler 480 System Gene Scanning Assay. The primers used, which were specific for the BRAF V600E mutation, were designed using Primer3 free software. The forward and reverse primer sequences were 5′-CATAATGCTTGCTCTGATAGGAAA-3′ and 5′-TCAGCACATCTCAGGGCCAAA-3′, respectively. All the primers were produced with standard molecular biology quality (Protech Technology Enterprise Co., Ltd., Taipei, Taiwan). RAS mutations were identified through direct DNA sequencing, the procedure for which was described in detail in our previous study [17]. Both KRAS and NRAS mutation statuses were examined in the patients. The presence of a deficient mismatch repair gene (dMMR) was determined by immunohistochemical staining of CRC tissue specimens. Loss of at least one mismatch repair protein (MLH-1, MSH-2, MSH-6, or PMS-2) was deemed indicative of the presence of dMMR [18].

2.3. Systemic Treatment and Outcome Assessment

In this case series, all eligible patients received the triplet regimen, which comprised the BRAF inhibitor dabrafenib (Novartis Pharmaceuticals, Basel, Switzerland), the MEK inhibitor trametinib (Novartis Pharmaceuticals, Basel, Switzerland), and the anti-EGFR agent panitumumab (Amgen Inc., Thousand Oaks, CA, USA) or cetuximab (Merck Sharp & Dohme Corp., Kenilworth, NJ, USA), after progressive disease was treated with at least second-line systemic treatment, including chemotherapy plus target therapy. The dosages were as follows: dabrafenib, 150 mg orally, twice per day; trametinib, 2 mg orally, once per day; panitumumab, 6 mg/kg every two weeks intravenously; and cetuximab, 400 mg/m^2 loading, then 500 mg/m^2 biweekly, intravenously. The patients attended regular follow-up visits at outpatient clinics every 2 weeks to evaluate symptoms and adverse events by the visiting staff and study nurses. When the patients were hospitalized for treatment or any other reason, the visiting staff and study nurses would be informed to allow assessment. The adverse events were recorded and graded during each cycle based on the National Cancer Institute Common Terminology Criteria for Adverse Events (Version 4.3; http://ctep.cancer.gov/reporting/ctc.html). Symptomatic treatments were provided for milder (grade 1-2) adverse events without interruption of systemic therapy, and the triplet therapy would be temporarily withheld for more severe adverse events (grade 3). Triplet therapy was only resumed if the adverse events were not life-threatening, and the patient got substantial improvement. The treatment response was typically assessed after 8–12 weeks of treatment by computed tomography, magnetic resonance imaging, or positron emission tomography according to the criteria of the Response Evaluation Criteria in Solid Tumors (RECIST; version 1.1) [19]. The median follow-up period was 14.5 (range, 1–26) months.

The primary outcome of this study was PFS, and secondary outcomes were OS, response rate (RR), disease control rate (DCR), and adverse events (AEs) of treatment. PFS was defined as the time from the initiation of the triplet regimen to the first radiological progression or tumor-related death, whichever came first. OS was defined as the time from the initiation of the triplet regimen to death due to any cause. DCR was represented as the percentage of patients with complete response, partial response, or stable disease as their best response.

2.4. Statistics

SPSS (Version 20.0; SPSS, Chicago, IL, USA) was used for all data analyses. The continuous variables were compared with Wilcoxon's signed-rank test, and categorical variables were compared using the Chi-square test. The Kaplan–Meier method was used to calculate PFS and OS, and a log-rank test was used to compare time-to-event distributions by clinical and molecular factors. Statistical significance was set at $p < 0.05$.

3. Results

This section may be divided by subheadings. It should provide a concise and precise description of the experimental results, their interpretation, as well as the experimental conclusions that can be drawn.

3.1. Baseline Characteristics of Included Patients

This case series included nine patients (4 had primary tumors on the right side: 2 in ascending colon and 2 in transverse colon; and 5 on the left side: 3 in descending colon cancer and 2 in sigmoid colon) with BRAF V600E-mutated mCRC who underwent triplet therapy. Their baseline characteristics are displayed in Table 1. All patients had tumors with wild-type KRAS/NRAS and moderate to poor differentiation. dMMR was noted in two of the six analyzed patients. Most patients received panitumumab, dabrafenib, and trametinib as the triplet regimen, but one patient used cetuximab instead of panitumumab. In addition, triplet therapy was exclusively used as third-line or later treatment in all but one patient, for whom the therapy was initiated after the failure of first-line therapy. Most patients (77.7%) had liver metastases, and in nearly half of them (44.4%), at least three organs were involved at the time of treatment. No significant differences in baseline characteristics were observed between left-sided and right-sided mCRC.

Table 1. Baseline characteristics of all included patients with *BRAF*-mutated mCRC receiving triplet therapy, stratified by tumor sidedness.

Characteristic	All Patients (*N* = 9)	Right Side Tumor (*N* = 4)	Left Side Tumor (*N* = 5)	*p* Value
Gender (Male: Female)	4:5	3:1	1:4	0.099
Age (years) Median ± SD (range)	51 ± 14.4 (35–81)	52.5 ± 5.8 (45–59)	45 ± 19.7 (35–81)	0.730
BMI kg/m^2 Mean ± SD	22.7 ± 6.4	22.8 ± 3.5	22.6 ± 8.5	1.000
Histology Moderately differentiated Poorly differentiated	 7 (77.7%) 2 (22.2%)	 3 (75%) 1 (25%)	 4 (80%) 1 (20%)	0.858
Stage at triplet therapy 4A 4B 4C	 4 (44.4%) 3 (33.3%) 2 (22.2%)	 2 (50%) 1 (25%) 1 (25%)	 2 (40%) 2 (40%) 1 (20%)	0.894
Involvement of ≥3 organs	4 (44.4%)	2 (50%)	2 (40%)	0.764
Liver metastasis	7 (77.7%)	3 (75%)	4 (75%)	0.858
Primary tumor resection Complete resection Partial or no resection	 5 (55.5%) 4 (44.4%)	 2 (50%) 2 (50%)	 3 (60%) 2 (40%)	0.764
Baseline CEA > 5 µg/L	8 (88.8%)	3 (75%)	5 (100%)	0.236
Response Partial response Stable disease Progressive disease	 1 (11.1%) 2 (22.2%) 6 (66.6%)	 0 1 (25%) 3 (75%)	 1 (20%) 1 (20%) 3 (60%)	0.638
Responder Non-responder	1 (11.1%) 8 (88.8%)	0 4 (100%)	1 (20%) 4 (80%)	0.343
Disease control rate	3 (33.3%)	1 (25%)	2 (40%)	0.635

SD, standard deviation; CEA, carcinoembryonic antigen; BMI: body mass index

3.2. Response Rate and Survival Analysis

Among the patients who underwent triplet therapy, only one patient had a partial response, and another two had stable disease (Table 2). All other patients had disease progression despite treatment (RR, 11.1%; DCR, 33.3%). The median PFS and OS were 2.9 months and 7.4 months, respectively (Figure 1A,B). No specific clinical or molecular factors were found to be significantly associated with favorable DCR or OS. However, a trend toward improved OS was found in left-sided mCRC compared with right-sided disease (9.2 vs. 6.9 months, $p = 0.093$) and patients with disease control. Median survival was not reached for patients with partial response or stable disease, and the median OS was 5.2 months for those with progressive disease ($p = 0.069$, Figure 2). In one patient with initial partial response after triplet therapy, PFS time persisted for 26 months until the last follow-up. In two patients with stable disease after triplet therapy, one had disease progression 3 months later and died, and the other patient achieved a PFS of 19 months without further systemic treatment.

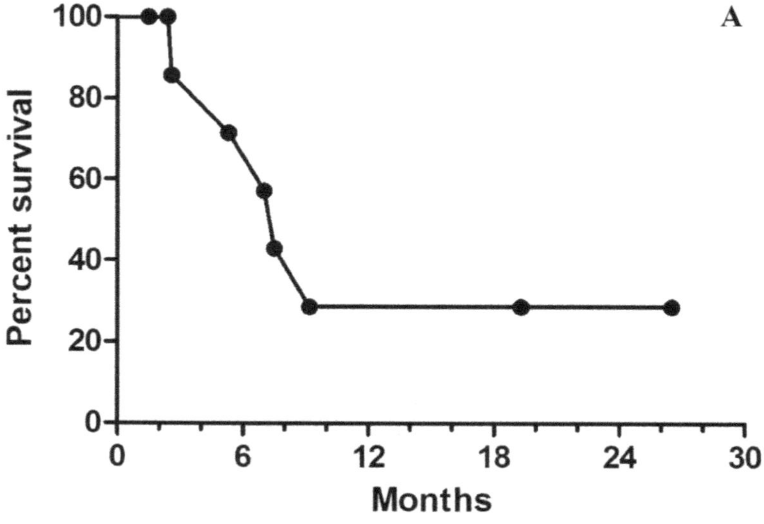

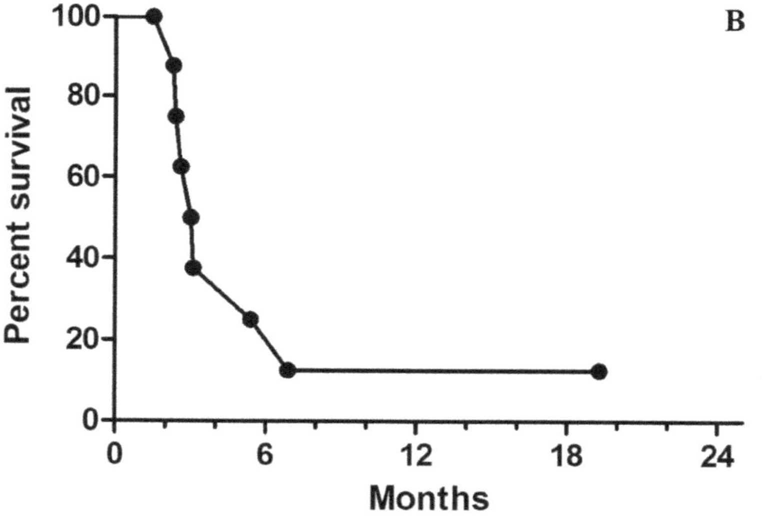

Figure 1. (**A**) Kaplan–Meier survival curves for median progression-free survival of 2.9 months for all nine patients; (**B**) Kaplan–Meier survival curves for median overall survival of 7.4 months for all nine patients.

Table 2. dMMR status, treatment responses, and survival of each patient with *BRAF*-mutated mCRC receiving triplet therapy.

	Age (Year) /Sex	Tumor Location	Primary Surgery	Metastasis Foci	dMMR	Best Response	PFS (Months)	OS (Months)
Patient 1	51, female	Right colon	No	Liver, lung, pancreas	ND	SD	6.9	7.5
Patient 2	45, female	Left colon	No	Liver, lung	No	PD	2.3	5.3
Patient 3	81, female	Left colon	R0 resection	Liver, lung	Yes	SD	19.3	19.3
Patient 4	41, male	Left colon	R0 resection	Liver, lung, adrenal gland, pancreas	No	PR	5.4	26.5
Patient 5	59, male	Right colon	R0 resection	Liver, peri-toneum, pancreas	ND	PD	3.1	7.0
Patient 6	45, male	Right colon	R0 resection	Liver, peri-toneum,	No	PD	1.5	1.5
Patient 7	54, male	Right colon	R1 resection	Peritoneum, Paraaortic lymph nodes	No	PD	2.6	2.6
Patient 8	35, female	Left colon	R0 resection	Peritoneum	Yes	PD	2.4	2.4
Patient 9	69, female	Left colon	No	Liver, lung, peri-toneum, bone	ND	PD	3.0	9.2

dMMR, deficiency of mismatch repair genes; ND, not done; PFS, progression-free survival; OS, overall survival; SD, stable disease; PR, partial response; PD, progressive disease; mCRC: metastatic colorectal cancer; R0: complete resection in gross with microscopically negative surgical margin; R1: complete resection in gross with microscopically positive surgical margin.

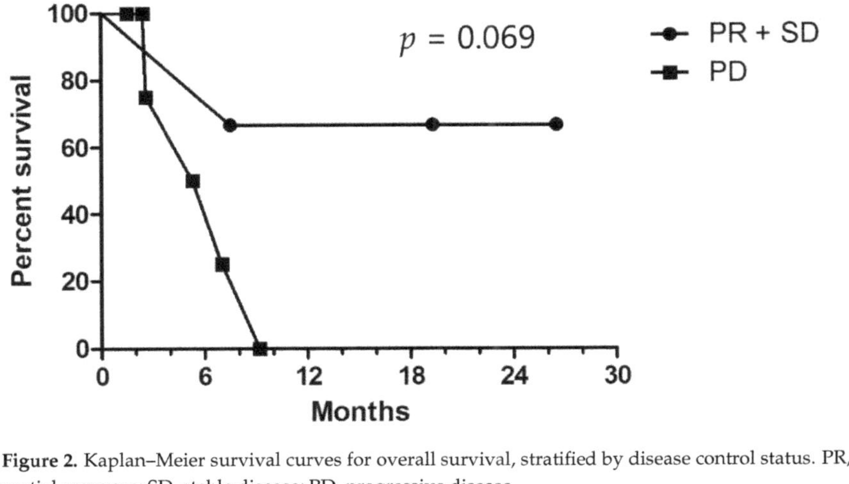

Figure 2. Kaplan–Meier survival curves for overall survival, stratified by disease control status. PR, partial response; SD, stable disease; PD, progressive disease.

3.3. Adverse Events

The adverse events in patients who received triplet therapy are summarized in Table 3. Triplet therapy was generally well-tolerated, and most adverse events were Grades 1–2. The most frequent adverse events were liver function abnormality (66.6%), hypertension (66.6%), and dermatitis (66.6%), followed by nausea (44.4%) and skin rash (44.4%). The most frequent severe events (Grade 3) were nausea (22%), hypertension (22%), dermatitis (22%), and diarrhea (11%). Of note, one patient developed blurred vision during the second month of triplet therapy, which gradually improved following the completion of systemic treatment and conservative management. No patient experienced grade 4 adverse events.

Table 3. Adverse events in all patients receiving triplet therapy for *BRAF*-mutated mCRC.

Adverse Events	Grade 1–2 (%)	Grade 3 (%) †	Any Grade (%)
Anemia	1 (11.1)	0 (0)	1 (11.1)
Neutropenia	0 (0)	0 (0)	0 (0)
Thrombocytopenia	0 (0)	0 (0)	0 (0)
Fatigue	0 (0)	0 (0)	0 (0)
Nausea	2 (22.2)	2 (22.2)	4 (44.4)
Vomiting	3 (33.3)	0 (0)	3 (33.3)
Hair loss	1 (11.1)	0 (0)	1 (11.1)
Abnormal liver function	6 (66.6)	0 (0)	6 (66.6)
Acute kidney injury	2 (22.2)	0 (0)	2 (22.2)
Hypertension	4 (44.4)	2 (22.2)	6 (66.6)
Diarrhea	1 (11.1)	1 (11.1)	2 (22.2)
Paresthesia	2 (22.2)	0 (0)	2 (22.2)
Skin rash	4 (44.4)	0 (0)	4 (44.4)
Dermatitis	4 (44.4)	2 (22.2)	6 (66.6)
Blurred vision	1 (11.1)	0 (0)	1 (11.1)

†: No patient had grade 4 adverse event in the study.

4. Discussion

In this case series, we demonstrated the real-world experience of using triplet therapy for BRAF-mutated mCRC as later lines of salvage therapy in Asian patients. Our findings suggest that triplet therapy appears to be well-tolerated and patients with initial disease control and longer PFS might gain considerable survival benefit, although most patients in our study still experienced disease progression.

The clinical efficacy of triplet therapy in BRAF-mutated mCRC has been demonstrated in two large clinical trials by Corcoran et al. [14] and Kopetz et al. (the BEACON trial) [11], and further trials are ongoing [15,20]. The trial by Corcoran et al. was a phase I trial using dabrafenib, panitumumab, and trametinib as triplet therapy, which was in line with our study's regimen. Triplet therapy resulted in a 21% RR, and median PFS and OS were 4.2 and 9.1 months, respectively. Nevertheless, the BEACON trial showed that triplet therapy (encorafenib, binimetinib, and cetuximab) had a 26% RR, and median PFS and OS were 4.3 months and 9.0 months, respectively; by contrast, the control group had only a 2% RR, and median PFS and OS were 1.5 months and 5.4 months, respectively. Of note, these two trials included a considerable portion of patients who failed to respond to first-line treatment; by contrast, in the current study, most patients previously underwent at least second-line systemic treatment. Although a direct comparison between our study and those mentioned above was not possible, the RR and survival in the current study seem acceptable.

Another compelling question is whether primary tumor location affects the outcome in BRAF-mutated mCRC treated with triplet therapy. Although the role of tumor location in the prognosis of BRAF-mutated mCRC remains controversial [16,21], it may have some impact with the concomitant use of target therapy such as bevacizumab or cetuximab [22]. Several studies have demonstrated that first-line bevacizumab plus chemotherapy resulted in a superior prognosis for right-sided BRAF-mutated mCRC [16,23–25]. Conversely, left-sided mCRC had more favorable outcomes when treated with anti-EGFR agents than did right-sided tumors [22,26], which was consistent with our observation. Further studies exploring the impact of the primary tumor side on the prognostic outcomes of BRAF-mutated mCRC treated with anti-EGFR agents may be quite valuable.

Several factors may have contributed to the discrepancies related to treatment response and survival between this study and others. First, our study included mostly patients who underwent two or more prior systemic treatments, a factor that has been found to be associated with a worse RR [27,28]. Second, the difference between clinical trials and real-world practice may lead to some bias in objective evaluation. Other clinical factors, such as the presence of dMMR [29], differences in ethnicity, and different regimens, as well as the genetic alteration patterns of the BRAF mutation [30], might also influence the outcomes. These factors warrant a higher case enrollment and detailed analysis to clarify the best candidates for triplet therapy as salvage therapy among BRAF-mutated mCRC patients.

Regarding adverse events with triplet therapy, the most common AEs, including gastrointestinal and dermatologic disorders, were similar to those in previous clinical trials [11,14]. Our study did not observe any cases that required dose escalation or discontinuation due to side effects; one patient developed blurred vision, but the cycle of triplet therapy was maintained after two weeks until this symptom subsided. A previous study demonstrated that MEK inhibitors can induce retinopathy [31]. The onset is typically rapid in the first week of treatment but often resolves gradually, even without drug interruption. Thus, although this unique adverse event must be carefully monitored, it typically does not cause a serious sequela.

To the best of our knowledge, this is the first real-world study of triplet therapy in BRAF-mutated mCRC in Asian patients. In this study, we demonstrated an acceptable safety profile for triplet therapy, and we expect prolonged survival when initial disease control is obtained, even with two or more failures of prior systemic treatments. However, the limited case number precluded a robust subgroup analysis, and more data are necessary to explore the predictive factors of the prognosis of triplet therapy for BRAF-mutated mCRC in real-world practice.

In summary, this single-center case series demonstrated that triplet therapy with BRAF and MEK inhibitors and an anti-EGFR agent had an acceptable safety profile and reasonable efficacy for BRAF-mutated mCRC. Further studies enrolling more patients are needed to identify potential treatment responses and improve the efficacy of the treatment regimen.

Author Contributions: Conceptualization, J.-H.Y., Y.-P.L. and J.-Y.W.; methodology, H.-L.T., Y.-C.C. and C.-C.L.; formal analysis, J.-H.Y. and P.-J.C.; investigation, W.-C.S. and C.-W.H.; data curation, C.-W.H. and T.-K.C.; project administration, H.-L.T.; funding acquisition, J.-Y.W.; writing—original draft preparation, J.-H.Y.; writing—review and editing, J.-Y.W. and Y.-P.L.; supervision, J.-Y.W. All authors have read and agreed to the published version of the manuscript.

Funding: This work was supported by grants through funding from the Ministry of Science and Technology (MOST 109-2314-B-037-035, MOST 109-2314-B-037-040, MOST 109-2314-B-037-046-MY3, MOST110-2314-B-037-097) and the Ministry of Health and Welfare (MOHW109-TDU-B-212-134026, MOHW109-TDU-B-212-114006, MOHW110-TDU-B-212-1140026) and funded by the health and welfare surcharge of on tobacco products, and the Kaohsiung Medical University Hospital (KMUH109-9R32, KMUH109-9R33, KMUH109-9R34, KMUH109-9M30, KMUH109-9M31, KMUH109-9M32, KMUH109-9M33, KMUHSA10903, KMUHSA11013, KMUH-DK(C)110010, KMUH-DK(B)110004-3) and KMU Center for Cancer Research (KMU-TC109A04-1) as well as and a KMU Center for Liquid Biopsy and Cohort Research Center Grant (KMU-TC109B05), Kaohsiung Medical University. In addition, this study was supported by the Grant of Taiwan Precision Medicine Initiative, Academia Sinica, Taiwan, R.O.C.

Institutional Review Board Statement: The authors are accountable for all aspects of the work in ensuring that questions related to the accuracy or integrity of any part of the work are appropriately investigated and resolved. The study was conducted in accordance with the Declaration of Helsinki (as revised in 2013). The study was approved by institutional/regional/national ethics/committee/ethics board of [KMUHIRB-2012-03-02(II)].

Informed Consent Statement: Individual consent for this retrospective analysis was waived.

Data Availability Statement: The datasets used and/or analysed during the current study are available from the corresponding author on reasonable request.

Conflicts of Interest: All authors have completed the ICMJE uniform disclosure form. The authors have no conflict of interest to declare.

References

1. Biller, L.H.; Schrag, D. Diagnosis and Treatment of Metastatic Colorectal Cancer: A Review. *JAMA* **2021**, *325*, 669–685. [CrossRef] [PubMed]
2. Siegel, R.L.; Miller, K.D.; Fuchs, H.E.; Jemal, A. Cancer Statistics, 2021. *CA Cancer J. Clin.* **2021**, *71*, 7–33. [CrossRef]
3. Dahabreh, I.J.; Terasawa, T.; Castaldi, P.J.; Trikalinos, T.A. Systematic review: Anti-epidermal growth factor receptor treatment effect modification by KRAS mutations in advanced colorectal cancer. *Ann. Intern. Med.* **2011**, *154*, 37–49. [CrossRef] [PubMed]
4. Tougeron, D.; Lecomte, T.; Pagès, J.C.; Villalva, C.; Collin, C.; Ferru, A.; Tourani, J.M.; Silvain, C.; Levillain, P.; Karayan-Tapon, L. Effect of low-frequency KRAS mutations on the response to anti-EGFR therapy in metastatic colorectal cancer. *Ann. Oncol. Off. J. Eur. Soc. Med. Oncol.* **2013**, *24*, 1267–1273. [CrossRef] [PubMed]
5. André, T.; Shiu, K.-K.; Kim, T.W.; Jensen, B.V.; Jensen, L.H.; Punt, C.; Smith, D.; Garcia-Carbonero, R.; Benavides, M.; Gibbs, P.; et al. Pembrolizumab in Microsatellite-Instability-High Advanced Colorectal Cancer. *N. Engl. J. Med.* **2020**, *383*, 2207–2218. [CrossRef] [PubMed]
6. Nazemalhosseini Mojarad, E.; Kashfi, S.M.; Mirtalebi, H.; Taleghani, M.Y.; Azimzadeh, P.; Savabkar, S.; Pourhoseingholi, M.A.; Jalaeikhoo, H.; Aghdaei, H.A.; Kuppen, P.J.K.; et al. Low Level of Microsatellite Instability Correlates with Poor Clinical Prognosis in Stage II Colorectal Cancer Patients. *J. Oncol.* **2016**, *2016*, 2196703.
7. Yuan, Z.X.; Wang, X.Y.; Qin, Q.Y.; Chen, D.F.; Zhong, Q.H.; Wang, L.; Wang, J.P. The prognostic role of BRAF mutation in metastatic colorectal cancer receiving anti-EGFR monoclonal antibodies: A meta-analysis. *PLoS ONE* **2013**, *8*, e65995. [CrossRef]
8. Chu, J.E.; Johnson, B.; Kugathasan, L.; Morris, V.K.; Raghav, K.; Swanson, L.; Lim, H.J.; Renouf, D.J.; Gill, S.; Wolber, R.; et al. Population-based Screening for BRAF (V600E) in Metastatic Colorectal Cancer Reveals Increased Prevalence and Poor Prognosis. *Clin. Cancer Res. Off. J. Am. Assoc. Cancer Res.* **2020**, *26*, 4599–4605. [CrossRef]
9. Modest, D.P.; Ricard, I.; Heinemann, V.; Hegewisch-Becker, S.; Schmiegel, W.; Porschen, R.; Stintzing, S.; Graeven, U.; Arnold, D.; von Weikersthal, L.F.; et al. Outcome according to KRAS-, NRAS- and BRAF-mutation as well as KRAS mutation variants: Pooled analysis of five randomized trials in metastatic colorectal cancer by the AIO colorectal cancer study group. *Ann. Oncol. Off. J. Eur. Soc. Med. Oncol.* **2016**, *27*, 1746–1753. [CrossRef]
10. Gavin, P.G.; Colangelo, L.H.; Fumagalli, D.; Tanaka, N.; Remillard, M.Y.; Yothers, G.; Kim, C.; Taniyama, Y.; Kim, S.I.; Choi, H.J.; et al. Mutation profiling and microsatellite instability in stage II and III colon cancer: An assessment of their prognostic and oxaliplatin predictive value. *Clin. Cancer Res. Off. J. Am. Assoc. Cancer Res.* **2012**, *18*, 6531–6541. [CrossRef]
11. Kopetz, S.; Grothey, A.; Yaeger, R.; Van Cutsem, E.; Desai, J.; Yoshino, T.; Wasan, H.; Ciardiello, F.; Loupakis, F.; Hong, Y.S.; et al. Encorafenib, Binimetinib, and Cetuximab in BRAF V600E-Mutated Colorectal Cancer. *N. Engl. J. Med.* **2019**, *381*, 1632–1643. [CrossRef]
12. Corcoran, R.B.; Atreya, C.E.; Falchook, G.S.; Kwak, E.L.; Ryan, D.P.; Bendell, J.C.; Hamid, O.; Messersmith, W.A.; Daud, A.; Kurzrock, R.; et al. Combined BRAF and MEK Inhibition with Dabrafenib and Trametinib in BRAF V600-Mutant Colorectal Cancer. *J. Clin. Oncol. Off. J. Am. Soc. Clin. Oncol.* **2015**, *33*, 4023–4031. [CrossRef]
13. Yaeger, R.; Cercek, A.; O'Reilly, E.M.; Reidy, D.L.; Kemeny, N.; Wolinsky, T.; Capanu, M.; Gollub, M.J.; Rosen, N.; Berger, M.F.; et al. Pilot trial of combined BRAF and EGFR inhibition in BRAF-mutant metastatic colorectal cancer patients. *Clin. Cancer Res. Off. J. Am. Assoc. Cancer Res.* **2015**, *21*, 1313–1320. [CrossRef] [PubMed]
14. Corcoran, R.B.; André, T.; Atreya, C.E.; Schellens, J.H.M.; Yoshino, T.; Bendell, J.C.; Hollebecque, A.; McRee, A.J.; Siena, S.; Middleton, G.; et al. Combined BRAF, EGFR, and MEK Inhibition in Patients with BRAF(V600E)-Mutant Colorectal Cancer. *Cancer Discov.* **2018**, *8*, 428–443. [CrossRef]
15. Grothey, A.; Tabernero, J.; Taieb, J.; Yaeger, R.; Yoshino, T.; Maiello, E.; Fernandez, E.E.; Casado, A.R.; Ross, P.; André, T.; et al. LBA-5 ANCHOR CRC: A single-arm, phase 2 study of encorafenib, binimetinib plus cetuximab in previously untreated BRAF V600E-mutant metastatic colorectal cancer. *Ann. Oncol.* **2020**, *31*, S242–S243. [CrossRef]

16. Hsieh, Y.-C.; Chang, T.-K.; Su, W.-C.; Huang, C.-W.; Tsai, H.-L.; Chen, Y.-C.; Li, C.-C.; Chen, P.-J.; Yin, T.-C.; Ma, C.-J.; et al. UGT1A1 Polymorphism for Irinotecan Dose Escalation in Patients with BRAF-Mutated Metastatic Colorectal Cancer Treated with First-Line Bevacizumab and FOLFIRI. *J. Oncol.* **2021**, *2021*, 6686517. [CrossRef]
17. Chuang, S.-C.; Huang, C.-W.; Chen, Y.-T.; Ma, C.-J.; Tsai, H.-L.; Chang, T.-K.; Su, W.-C.; Hsu, W.-H.; Kuo, C.-H.; Wang, J.-Y. Effect of KRAS and NRAS mutations on the prognosis of patients with synchronous metastatic colorectal cancer presenting with liver-only and lung-only metastases. *Oncol. Lett.* **2020**, *20*, 2119–2130. [CrossRef] [PubMed]
18. Weissman, S.M.; Bellcross, C.; Bittner, C.C.; Freivogel, M.E.; Haidle, J.L.; Kaurah, P.; Leininger, A.; Palaniappan, S.; Steenblock, K.; Vu, T.M.; et al. Genetic counseling considerations in the evaluation of families for Lynch syndrome—A review. *J. Genet. Couns.* **2011**, *20*, 5–19. [CrossRef]
19. Therasse, P.; Arbuck, S.G.; Eisenhauer, E.A.; Wanders, J.; Kaplan, R.S.; Rubinstein, L.; Verweij, J.; Van Glabbeke, M.; Van Oosterom, A.T.; Christian, M.C.; et al. New guidelines to evaluate the response to treatment in solid tumors. European Organization for Research and Treatment of Cancer, National Cancer Institute of the United States, National Cancer Institute of Canada. *J. Natl. Cancer Inst.* **2000**, *92*, 205–216. [CrossRef] [PubMed]
20. Kotani, D.; Bando, H.; Masuishi, T.; Komatsu, Y.; Yamaguchi, K.; Nakajima, T.; Satoh, T.; Nishina, T.; Esaki, T.; Nomura, S.; et al. BIG BANG study: A multicenter phase II study of the MEK inhibitor binimetinib + BRAF inhibitor encorafenib + anti-EGFR antibody cetuximab in patients with BRAF non-V600E mutated metastatic colorectal cancer (EPOC 1703). *Ann. Oncol.* **2018**, *29*, viii201–viii202. [CrossRef]
21. Loupakis, F.; Intini, R.; Cremolini, C.; Orlandi, A.; Sartore-Bianchi, A.; Pietrantonio, F.; Pella, N.; Spallanzani, A.; Dell'Aquila, E.; Scartozzi, M.; et al. A validated prognostic classifier for V600EBRAF-mutated metastatic colorectal cancer: The 'BRAF BeCool' study. *Eur. J. Cancer* **2019**, *118*, 121–130. [CrossRef] [PubMed]
22. Brulé, S.; Jonker, D.; Karapetis, C.; O'Callaghan, C.; Moore, M.; Wong, R.; Tebbutt, N.; Underhill, C.; Yip, D.; Zalcberg, J.; et al. Location of colon cancer (right-sided versus left-sided) as a prognostic factor and a predictor of benefit from cetuximab in NCIC CO.17. *Eur J. Cancer* **2015**, *51*, 1405–1414. [CrossRef]
23. Ulivi, P.; Scarpi, E.; Chiadini, E.; Marisi, G.; Valgiusti, M.; Capelli, L.; Gardini, A.C.; Monti, M.; Ruscelli, S.; Frassineti, G.L.; et al. Right- vs. Left-Sided Metastatic Colorectal Cancer: Differences in Tumor Biology and Bevacizumab Efficacy. *Int. J. Mol. Sci.* **2017**, *18*, 1240. [CrossRef] [PubMed]
24. Venook, A.P.; Niedzwiecki, D.; Innocenti, F.; Fruth, B.; Greene, C.; O'Neil, B.H.; Shaw, J.E.; Atkins, J.N.; Horvath, L.E.; Polite, B.N.; et al. Impact of primary (1°) tumor location on overall survival (OS) and progression-free survival (PFS) in patients (pts) with metastatic colorectal cancer (mCRC): Analysis of CALGB/SWOG 80405 (Alliance). *J. Clin. Oncol.* **2016**, *34* (Suppl. 15), 3504. [CrossRef]
25. Stintzing, S.; Heinrich, K.; Tougeron, D.; Modest, D.P.; Schwaner, I.; Euker, J.; Pihusch, R.; Stauch, M.; Kaiser, F.; Kahl, C.; et al. Randomized study to investigate FOLFOXIRI plus either bevacizumab or cetuximab as first-line treatment of BRAF V600E-mutant mCRC: The phase-II FIRE-4.5 study (AIO KRK-0116). *J. Clin. Oncol.* **2021**, *39* (Suppl. 15), 3502. [CrossRef]
26. Yaeger, R.; Kotani, D.; Mondaca, S.; Parikh, A.R.; Bando, H.; Van Seventer, E.E.; Taniguchi, H.; Zhao, H.; Thant, C.N.; De Stanchina, E.; et al. Response to Anti-EGFR Therapy in Patients with BRAF non-V600–Mutant Metastatic Colorectal Cancer. *Clin. Cancer Res.* **2019**, *25*, 7089–7097. [CrossRef]
27. Morris, V.; Overman, M.J.; Jiang, Z.-Q.; Garrett, C.; Agarwal, S.; Eng, C.; Kee, B.; Fogelman, D.; Dasari, A.; Wolff, R.; et al. Progression-free survival remains poor over sequential lines of systemic therapy in patients with BRAF-mutated colorectal cancer. *Clin. Colorectal Cancer* **2014**, *13*, 164–171. [CrossRef]
28. Van Cutsem, E.; Huijberts, S.; Grothey, A.; Yaeger, R.; Cuyle, P.J.; Elez, E.; Fakih, M.; Montagut, C.; Peeters, M.; Yoshino, T.; et al. Binimetinib, Encorafenib, and Cetuximab Triplet Therapy for Patients with BRAF V600E-Mutant Metastatic Colorectal Cancer: Safety Lead-in Results from the Phase III BEACON Colorectal Cancer Study. *J. Clin. Oncol. Off. J. Am. Soc. Clin. Oncol.* **2019**, *37*, 1460–1469. [CrossRef]
29. Fanelli, G.N.; Pozzo, C.A.D.; Depetris, I.; Schirripa, M.; Brignola, S.; Biason, P.; Balistreri, M.; Santo, L.D.; Lonardi, S.; Munari, G.; et al. The heterogeneous clinical and pathological landscapes of metastatic Braf-mutated colorectal cancer. *Cancer Cell Int.* **2020**, *20*, 30. [CrossRef]
30. Middleton, G.; Yang, Y.; Campbell, C.D.; André, T.; Atreya, C.E.; Schellens, J.H.M.; Yoshino, T.; Bendell, J.C.; Hollebecque, A.; McRee, A.J.; et al. BRAF-Mutant Transcriptional Subtypes Predict Outcome of Combined BRAF, MEK, and EGFR Blockade with Dabrafenib, Trametinib, and Panitumumab in Patients with Colorectal Cancer. *Clin. Cancer Res.* **2020**, *26*, 2466. [CrossRef]
31. Booth, A.E.C.; Hopkins, A.M.; Rowland, A.; Kichenadasse, G.; Smith, J.R.; Sorich, M.J. Risk factors for MEK-associated retinopathy in patients with advanced melanoma treated with combination BRAF and MEK inhibitor therapy. *Ther. Adv. Med. Oncol.* **2020**, *12*, 1758835920944359. [CrossRef] [PubMed]

medicina MDPI

Case Report

A Promising Role of TGF-β Pathway in Response to Regorafenib in Metastatic Colorectal Cancer: A Case Report

Simona De Summa [1,†], Katia Danza [1,†], Brunella Pilato [1], Giuseppina Matera [1], Rossella Fasano [2], Angela Calabrese [3], Rosanna Lacalamita [1], Nicola Silvestris [2,4], Stefania Tommasi [1], Antonella Argentiero [2,‡] and Oronzo Brunetti [2,*,‡]

1 Molecular Diagnostics and Pharmacogenetics Unit, IRCCS Istituto Tumori "Giovanni Paolo II", 70124 Bari, Italy; s.desumma@oncologico.bari.it (S.D.S.); danzakatia@gmail.com (K.D.); brunellapilato@gmail.com (B.P.); g.matera92@gmail.com (G.M.); rosanna.lacalamita@gmail.com (R.L.); s.tommasi@oncologico.bari.it (S.T.)
2 Medical Oncology Unit, IRCCS Istituto Tumori "Giovanni Paolo II" of Bari, 70124 Bari, Italy; rossella.fasano.93@gmail.com (R.F.); n.silvestris@oncologico.bari.it (N.S.); argentieroantonella@gmail.com (A.A.)
3 Radiology Unit, IRCCS Istituto Tumori "Giovanni Paolo II", 70124 Bari, Italy; acalabrese22@gmail.com
4 Department of Biomedical Sciences and Human Oncology, University of Bari "Aldo Moro", 70124 Bari, Italy
* Correspondence: dr.oronzo.brunetti@tiscali.it
† These authors contributed equally to this work.
‡ Last co-authors.

Citation: De Summa, S.; Danza, K.; Pilato, B.; Matera, G.; Fasano, R.; Calabrese, A.; Lacalamita, R.; Silvestris, N.; Tommasi, S.; Argentiero, A.; et al. A Promising Role of TGF-β Pathway in Response to Regorafenib in Metastatic Colorectal Cancer: A Case Report. *Medicina* **2021**, *57*, 1241. https://doi.org/10.3390/medicina57111241

Academic Editors: Antonio M. Scanu and Maria Rosaria De Miglio

Received: 7 October 2021
Accepted: 11 November 2021
Published: 13 November 2021

Publisher's Note: MDPI stays neutral with regard to jurisdictional claims in published maps and institutional affiliations.

Abstract: Colorectal cancer (CRC) is one of the most common cancer types around the world. The prognosis of patients with advanced diseases is still poor in spite of currently available therapeutic options. Regorafenib is an oral tyrosine kinase inhibitor (TKI) approved to treat refractory metastatic colorectal cancer (mCRC). We investigated Somatic mutations in several genes involved in immunological response and cancer progression in both long/short responder mCRC patients who underwent third-line therapy with regorafenib to identify predictive biomarkers of response using Ion Torrent PGM sequencing and bioinformatic tools. We found Somatic mutations in TGFBR1, TGFBR2, and TGFBR3 genes in primary tumor and metastases samples of long-responder patients. Furthermore, our bioinformatic results show that they were mainly enriched in immune response, cell junction, and cell adhesion in long responder patients, particularly in primary tumor and metastatic sites. These data suggest that the TGF-b pattern could be the leading actor of a prolonged response to this drug.

Keywords: colorectal cancer; TGF-b; regorafenib

1. Introduction

Colorectal cancer (CRC) is the third leading cause of cancer death globally, and its incidence is steadily rising in developing nations [1–3]. In recent years, many targeted therapeutic strategies have been proposed for metastatic colorectal cancer (mCRC) patients [4]. Regorafenib is an oral type II multi-kinase inhibitor that inhibits the activity of vascular endothelial growth factor receptor 1, 2, 3 (VEGFR-1, -2, -3), platelet-derived growth factor receptors, fibroblast growth factor receptors (FGFR), tyrosine kinase receptor with immunoglobulin-like and EGF-like domains 2 (TIE-2), and oncogenic receptor tyrosine kinases [5], showing an impact on angiogenesis and metastasis processes. Observational post-marketing studies have confirmed its efficacy in the face of a non-negligible toxicity profile [6,7]. It has been approved in the third or later lines of treatment for patients with chemorefractory mCRC, according to the results of two randomized phases III trials (CORRECT and CONCUR) [8,9]. Unlike anti-EGFR monoclonal antibodies for which well-defined molecular predictive factors are available [10], results for anti-angiogenic drugs are still inconclusive [11]. However, since its introduction into the clinical setting, single exceptional responders to this drug have been reported [12–15]. Thus far, identifying useful predictive factors to identify the best candidates for such a therapy represents a relevant clinical challenge.

A more remarkable progression-free survival (PFS) benefit for regorafenib has been observed in patients showing epithelial-mesenchymal transition (EMT) phenotype and higher TGF-β pathway activation [16]. TGF-β pathway promotes angiogenesis and EMT and inhibits the growth of epithelial and immune cells [17,18]. Loss of SMAD4, a tumor suppressor gene, disrupts R-SMAD-SMAD4 complexes in the canonical TGF-β signaling, leading to the deregulation of several SMAD4-related target genes, such as VEGF-A, VEGF-C, and β-catenin [19]. A strict and mutual regulation between vasculature normalization and immune activation has been described in the tumor microenvironment [20]. Consequently, in this study, we evaluate somatic mutations of several genes involved in immunological response and cancer progression, with the aim to identify, through a NGS platform, potential biomarkers of response to regorafenib in one very long and short responder mCRC patient.

2. Materials and Methods

2.1. Patients' Characteristics

The clinical history of the 58-year-old mCRC who presented a prolonged progression-free survival (PFS) (16 months) to third-line regorafenib has been previously published [21].

The control case was a 54-year-old female mCRC patient who received the third-line treatment with regorafenib with a PFS of 4 months. This study was approved by the Local Ethical Committee (Prot. N.709/CE). Both patients provided informed consent.

2.2. Sample Processing

All surgical samples were formalin-fixed and paraffin-embedded (FFPE). Tumor sections were cut from each FFPE block: one section was stained by hematoxylin/eosin to confirm and locate the tumor, and consecutive sections were used for immunohistochemistry and gene expression analyses.

2.2.1. DNA and RNA Extraction

Three to six FFPE tissue sections (6 μm thick) with adequate tumor cellularity, selected by a pathologist (>50%), were macro dissected and subjected to the QIAamp DNA FFPE Tissue Kit (Qiagen, Venlo, The Netherlands) for DNA isolation and the RNeasy FFPE Kit (Qiagen) for RNA isolation, according to the manufacturer's protocols. DNA was also isolated from blood samples using the QIAamp DNA Blood Midi Kit (Qiagen). DNA and RNA concentrations were measured using the Qubit 2.0 fluorometer (Thermo Fisher Scientific, Waltham, MA, USA).

2.2.2. Ion Torrent PGM Sequencing

The sequencing which has been used in the current study has been reported in our previous study [22]. Briefly, two custom panels have been designed through the Ion Ampliseq designer tool, one including the coding region of 41 genes to detect Somatic mutations and one to study the gene expression of 95 genes. Both include genes involved in immune regulation and inflammation. Variant calling and filtering have been described in [22]. In particular, somatic variants were called when matching the following conditions: DP > 50, VD > 20, and QUAL > 30. Furthermore, the Cancer hotspot Panel V2 has been used to identify druggable alterations. Briefly, the call set has been generated merging results from the Somatic High-Stringency Variant Caller plugin of the Torrent Suite and the Vardict [23] algorithm. Germline variants were filtered out using a pool of healthy controls. Annovar [24] was used to annotate variants functionally. The Oncoprint plot has been designed with the ComplexHeatmap R package [25].

2.2.3. MSI Analysis

MSI Analysis was performed by using Real-Time PCR (Easy PGX Diatech) to detect the microsatellite region instability in tumor samples. The analysis of 8 mononucleotide markers (BAT-25, BAT-26, NR-21, NR-22, NR-24, NR-27, CAT-25, and MONO-27) was based on the denaturation profile and compared to a positive control with a stable profile and followed manufacture instructions (EasyPGX® ready MSI cat.no. RT033).

The EasyPGX® Analysis Software was used to analyze all the melting temperatures and to generate a melting profile from each sample and the positive control. The software automatically calculates the status of global instability (MSS, MSI-L, MSI-H) from the number of unstable markers.

A tumor with high instability (H-MSI) has $\geq$2 unstable markers, and a tumor with L-MSI (low instability) has 1 unstable marker. For all L-MSI tumors, it is needed to repeat the analysis comparing the melting profile of the tumor tissue to the melting profile of normal tissue to exclude possible germline instability.

2.3. Protein-Protein Interaction (PPI) Network and Pathway Enrichment Analysis

Metascape (Available online: https://metascape.org/gp/index.html#/main/step1 (accessed on 10 July 2020)), an online resource, has been used to depict biological networks, including the interaction between mutated genes and to perform functional enrichments.

3. Results

3.1. Mutational Pattern

Through a custom targeted NGS panel including 41 genes, long-responder and short-responder samples were sequenced. In detail, long-responder samples included a primary tumor and two ovarian metachronous metastases. The control case included a primary tumor and one lung metastasis. The primary tumor and the first metastasis showed a distinct pattern of alterations (only CD276, ICAM1, and ARHGEF7 mutations are common) (Figure 1A). The second metastasis shared almost all mutations detected in the primary tumor: four alterations were detected in the first metastasis with 9 private mutations (Figure 1B). The mutational pattern of the short-responder in the primary and metastatic samples reflects a simple mechanism of clonal evolution (Figure 1C). Indeed, we observed that metastasis has the same alterations as the primary tumor samples with a further 9 private mutations. Microsatellite status has been checked in all samples, which were found to be stable (MSS).

3.2. Functional Enrichment and PPI Network

The global pathway enrichment is displayed as a heatmap (Figure 2). Terms related to immune response, cell junction, and cell adhesion molecules are the most enriched in long-responder patients, particularly in the primary tumor and second metastasis samples.

The PPI network, including proteins with relative genes which were found to be altered in the second metastatic samples, was built up. It can be observed that two sub-networks were identified, which were, in turn, functionally enriched (Figure 3A).

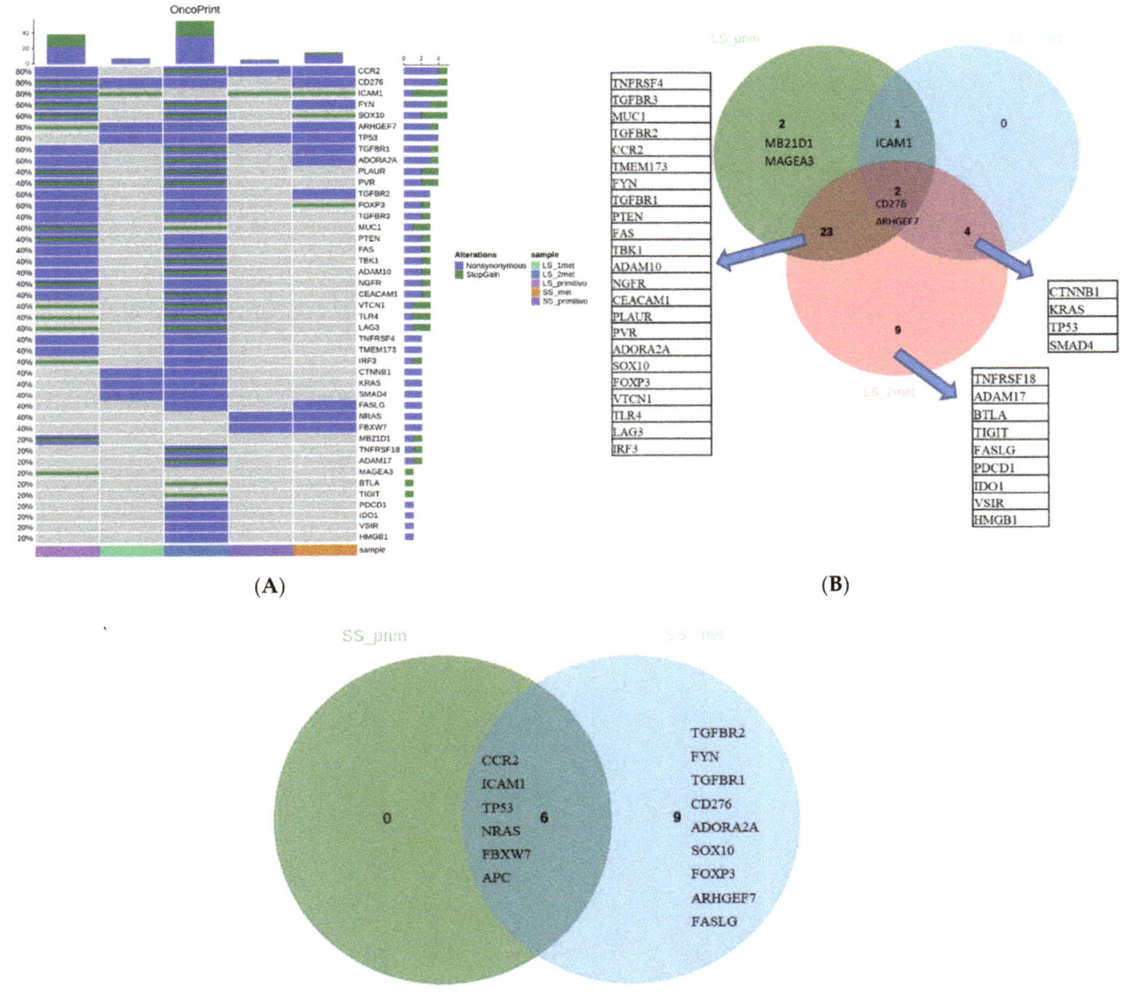

Figure 1. (**A**) Oncoprint including pathogenic alterations detected in the five analyzed samples. (**B**) Venn diagram of the alterations detected in the long survival samples (primary CRC and the two metastatic samples). (**C**) Venn diagram of the alterations detected in the short survival samples (primary CR Cand metastatic samples). LS_prim: primary tumor of the long survival case; LS_1met, LS_2met: metastatic samples of the long survival case; SS_prim: primary tumor of the short survival case; SS_met: metastatic sample of the short survival patient.

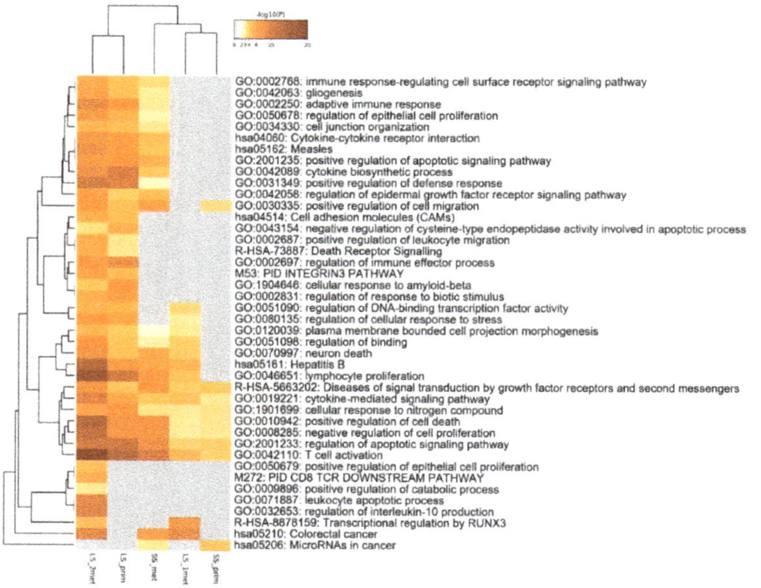

Figure 2. Global pathway enrichment of the detected alterations. LS_prim: primary tumor of the long survival case; LS_1met, LS_2met: metastatic samples of the long survival case; SS_prim: primary tumor of the short survival case; SS_met: metastatic sample of the short survival patient.

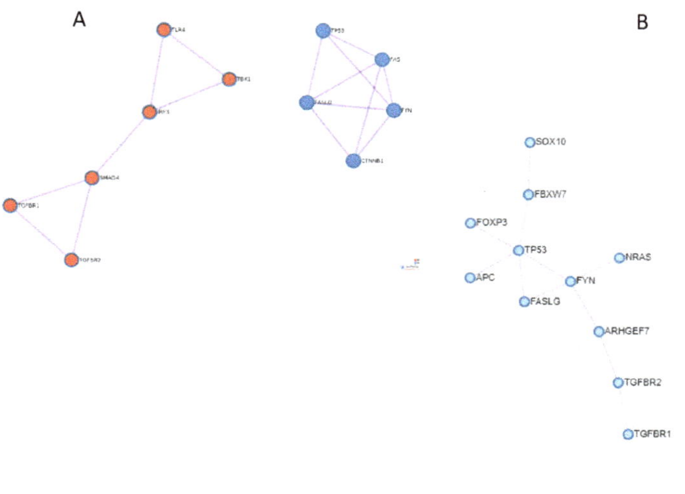

Figure 3. Interaction networks of the alterations detected in (**A**) long survival samples and (**B**) in short survival samples. In Table 1, significantly enriched terms are shown according to the sub-network color scale. PPI network was built up also for the short responder case (Figure 3B) and functionally enriched (Table 1).

Table 1. The significant genes enriched in the different pathways. In red are the enriched pathways for the network in Figure 3A; in blue is the sub-network in Figure 3B.

Category	GO	Description	LogP
KEGG Pathway	hsa05161	Hepatitis B	−10
Reactome Gene Sets	R-HSA-3304349	Loss of Function of SMAD2/3 in Cancer	−9.5
Reactome Gene Sets	R-HSA-3304351	Signaling by TGF-beta Receptor Complex in Cancer	−9.3
GO Biological Processes	GO:0045351	type I interferon biosynthetic process	−8.9
Reactome Gene Sets	R-HSA-936964	Activation of IRF3/IRF7 mediated by TBK1/IKK epsilon	−8.2
GO Biological Processes	GO:0003198	epithelial to mesenchymal transition involved in endocardial cushion formation	−8
GO Biological Processes	GO:0032727	positive regulation of interferon-alpha production	−7.8
Canonical Pathways	M185	PID ALK1 PATHWAY	−7.7
GO Biological Processes	GO:0003272	endocardial cushion formation	−7.6
GO Biological Processes	GO:0035666	TRIF-dependent toll-like receptor signaling pathway	−7.5
GO Biological Processes	GO:0032647	regulation of interferon-alpha production	−7.5
GO Biological Processes	GO:0032728	positive regulation of interferon-beta production	−7.4
GO Biological Processes	GO:0032607	interferon-alpha production	−7.4
Reactome Gene Sets	R-HSA-2173789	TGF-beta receptor signaling activates SMADs	−7.4
GO Biological Processes	GO:0002756	MyD88-independent toll-like receptor signaling pathway	−7.3
GO Biological Processes	GO:0060317	cardiac epithelial to mesenchymal transition	−7.2
GO Biological Processes	GO:0003203	endocardial cushion morphogenesis	−7.2
GO Biological Processes	GO:2000826	regulation of heart morphogenesis	−7
GO Biological Processes	GO:0060412	ventricular septum morphogenesis	−6.9
GO Biological Processes	GO:0003197	endocardial cushion development	−6.9
GO Biological Processes	GO:1901216	positive regulation of neuron death	−8.9
GO Biological Processes	GO:2001233	regulation of apoptotic signaling pathway	−8.9
KEGG Pathway	hsa05162	Measles	−8.4
GO Biological Processes	GO:0097190	apoptotic signaling pathway	−8
Reactome Gene Sets	R-HSA-109581	Apoptosis	−7.8
Reactome Gene Sets	R-HSA-5357801	Programmed Cell Death	−7.8
KEGG Pathway	hsa05205	Proteoglycans in cancer	−7.6
GO Biological Processes	GO:0010942	positive regulation of cell death	−7.6
GO Biological Processes	GO:0043523	regulation of neuron apoptotic process	−7.5
GO Biological Processes	GO:2001234	negative regulation of apoptotic signaling pathway	−7.4
GO Biological Processes	GO:0051402	neuron apoptotic process	−7.3
GO Biological Processes	GO:0070266	necroptotic process	−7.3
GO Biological Processes	GO:0097300	programmed necrotic cell death	−7.1
GO Biological Processes	GO:0043525	positive regulation of neuron apoptotic process	−6.9
GO Biological Processes	GO:1901214	regulation of neuron death	−6.9
GO Biological Processes	GO:0070265	necrotic cell death	−6.8
GO Biological Processes	GO:0070997	neuron death	−6.7
KEGG Pathway	hsa01524	Platinum drug resistance	−6.6
KEGG Pathway	hsa05200	Pathways in cancer	−6.5
GO Biological Processes	GO:2001237	negative regulation of extrinsic apoptotic signaling pathway	−6.1

4. Discussion

Despite improvements in the management of mCRC, drug resistance remains a clinical challenge in the advanced stage. Regorafenib was the first approved multikinase inhibitor with survival benefits in unselected mCRC patients who had exhausted current standard therapies [8,9]. Regorafenib inhibits the activity of several protein kinases active in the regulation of angiogenesis, oncogenesis, and in the modulation of the tumor microenvironment. Although several clinical and biological parameters have been investigated, there are no useful predictive markers for regorafenib treatment [26,27]. In this study, we explored Somatic mutations of genes involved in immunological and inflammation response in one very long-responder and one short-responder mCRC patient to regorafenib. Recently, an immune profile that correlates with the outcome in mCRC patients treated with regorafenib has been reported, suggesting a cytokine signature able to discriminate patients who might derive a benefit from regorafenib treatment [28]. In particular, the plasma basal level of proteins TNF-α and TGF-β before treatment might be useful to identify mCRC patients that do not benefit from regorafenib and show the progression of the disease. According to these results, our data show most mutated genes involved in TGF-β signaling in long-responder mCRC patients to regorafenib therapy.

In particular, Somatic mutations in TGFBR1, TGFBR2, and TGFBR3 genes were found in the primary tumor and metastatic samples of our long-responder patient. TGF- β was identified as a major signaling pathway in CRC invasion and metastasis. Its activation generally promotes CRC invasion and metastasis through EMT, whereas it suppresses cancer immunity in the tumor microenvironment [29]. The relevance of TGF-β signaling in the acquisition of an invasive phenotype was also demonstrated by our PPI subnetwork obtained by protein products of altered genes found in the last metastasis of the long-responder mCRC patient. This subnetwork was enriched by terms related to "Signaling by TGF-beta Receptor Complex in Cancer" and "TGF-beta receptor signaling activates SMADs". No similar terms were observed by the enrichment of the PPI network highlighted by protein products of altered genes found in metastases of the short-responder.

Interestingly, in a study, researchers showed a deleterious mutation in the SMAD4 gene in the long-responder metastatic patient [30]. Martinelli et al. [16] reported a greater PFS benefit for regorafenib therapy in patients with SMAD4 gene mutation characterized by the activation of TGFβ signaling and upregulation of an EMT pathway [31]. Authors observed mutation in SMAD4 in two long-responder patients, suggesting a key role of this gene in regorafenib response [16]. Down-regulation or mutation of SMAD4 underlies a more rapid protein degradation, leading to pancreatic cancer cell cycle arrest and apoptosis [32]. A key role of this member of TGFβ signaling has also been reported in CRC cells, in which deletion of SMAD4 decreased the number of TAMs in the tumor microenvironment, contributing to unfavorable prognoses [19].

Regorafenib appears to participate in the immune system with tumor interaction in different ways, including inhibition of tyrosine kinase receptor CSF1R, which is involved in macrophage proliferation [33]. Recently, a strict and mutual regulation between vasculature normalization and immune activation has been described in the tumor microenvironment [20]. In conclusion, we can hypothesize that the TGF-b pattern could be the leading actor of a longer response; however, all our results should be better explored in a larger cohort.

Author Contributions: S.D.S. and K.D., the first authors of the manuscript, designed the project, performed the experiment, analyzed the data, and wrote the initial version of the manuscript. R.F. and A.C. collected the samples and provided comments. B.P., G.M., R.L. and S.T. performed the surgical procedures. N.S. and A.A. interpreted the results, helped in data categorization, and critically reviewed the manuscript. O.B., the corresponding author of the manuscript, supervised the project and revised the main text of the manuscript. All authors have read and agreed to the published version of the manuscript.

Funding: This research received no external funding.

Institutional Review Board Statement: This study was approved by the Ethics Committee of Cancer institute of Bari (Prot. n° 209/CE).

Informed Consent Statement: All participants signed written informed consent before the experiment.

Data Availability Statement: Not available.

Conflicts of Interest: The authors declare that there is no conflict of interest.

References

1. Rumpold, H.; Niedersüß-Beke, D.; Heiler, C.; Falch, D.; Wundsam, H.V.; Metz-Gercek, S.; Piringer, G.; Thaler, J. Prediction of mortality in metastatic colorectal cancer in a real-life population: A multicenter explorative analysis. *BMC Cancer* **2020**, *20*, 1–9. [CrossRef] [PubMed]
2. Derakhshani, A.; Hashemzadeh, S.; Asadzadeh, Z.; Shadbad, M.A.; Rasibonab, F.; Safarpour, H.; Jafarlou, V.; Solimando, A.G.; Racanelli, V.; Singh, P.K.; et al. Cytotoxic T-Lymphocyte Antigen-4 in Colorectal Cancer: Another Therapeutic Side of Capecitabine. *Cancers* **2021**, *13*, 2141. [CrossRef]
3. Asadzadeh, Z.; Mansoori, B.; Mohammadi, A.; Kazemi, T.; Mokhtarzadeh, A.; Shanehbandi, D.; Hemmat, N.; Derakhshani, A.; Brunetti, O.; Safaei, S.; et al. The combination effect of Prominin1 (CD133) suppression and Oxaliplatin treatment in colorectal cancer therapy. *Biomed. Pharmacother. Biomed. Pharmacother.* **2021**, *137*, 111364. [CrossRef] [PubMed]
4. Xie, Y.-H.; Chen, Y.-X.; Fang, J.-Y. Comprehensive review of targeted therapy for colorectal cancer. *Signal Transduct. Target. Ther.* **2020**, *5*, 1–30. [CrossRef] [PubMed]
5. Wilhelm, S.M.; Dumas, J.; Adnane, L.; Lynch, M.; Carter, C.A.; Schütz, G.; Thierauch, K.H.; Zopf, D. Regorafenib (BAY 73-4506): A new oral multikinase inhibitor of angiogenic, stromal and oncogenic receptor tyrosine kinases with potent preclinical antitumor activity. *Int. J. Cancer* **2011**, *129*, 245–255. [CrossRef] [PubMed]
6. O'Connor, J.M.; Öhler, L.; Scheithauer, W.; Metges, J.-P.; Dourthe, L.-M.; de Groot, J.W.; Thaler, J.; Yeh, K.-H.; Lin, J.-K.; Falcone, A. Real-world dosing of regorafenib in metastatic colorectal cancer (mCRC): Interim analysis from the prospective, observational CORRELATE study. *Liver* **2017**, *260*, 52. [CrossRef]
7. Tougeron, D.; Desseigne, F.; Etienne, P.; Dourthe, L.; Mineur, L.; Paule, B.; Hollebecque, A.; Tresch, E.; Spaeth, D.; Michel, P. REBECCA: A large cohort study of regorafenib (REG) in the real-life setting in patients (pts) previously treated for metastatic colorectal cancer (mCRC). *Ann. Oncol.* **2014**, *25*, iv205. [CrossRef]
8. Grothey, A.; Van Cutsem, E.; Sobrero, A.; Siena, S.; Falcone, A.; Ychou, M.; Humblet, Y.; Bouché, O.; Mineur, L.; Barone, C. Regorafenib monotherapy for previously treated metastatic colorectal cancer (CORRECT): An international, multicentre, randomised, placebo-controlled, phase 3 trial. *Lancet* **2013**, *381*, 303–312. [CrossRef]
9. Li, J.; Qin, S.; Xu, R.; Yau, T.C.; Ma, B.; Pan, H.; Xu, J.; Bai, Y.; Chi, Y.; Wang, L. Regorafenib plus best supportive care versus placebo plus best supportive care in Asian patients with previously treated metastatic colorectal cancer (CONCUR): A randomised, double-blind, placebo-controlled, phase 3 trial. *Lancet Oncol.* **2015**, *16*, 619–629. [CrossRef]
10. Silvestris, N.; Tommasi, S.; Santini, D.; Russo, A.; Simone, G.; Petriella, D.; Maiello, E.; Tonini, G.; Colucci, G. KRAS mutations and sensitivity to anti-EGFR monoclonal antibodies in metastatic colorectal carcinoma: An open issue. *Expert Opin. Biol. Ther.* **2009**, *9*, 565–577. [CrossRef]
11. Hegde, P.S.; Wallin, J.J.; Mancao, C. Predictive markers of anti-VEGF and emerging role of angiogenesis inhibitors as immunotherapeutics. In *Seminars in Cancer Biology*; Academic Press: Cambridge, MA, USA, 2018; pp. 117–124.
12. Rosati, G.; Del Gaudio, N.; Scarano, E.; Cifarelli, R.A.; Altucci, L.; Bilancia, D. Unexpected and durable response with regorafenib in a metastatic colorectal cancer patient without KDR mutation: A case report. *Medicine* **2018**, *97*, e11178. [CrossRef] [PubMed]
13. Callebout, E.; Ribeiro, S.M.; Laurent, S.; De Man, M.; Ferdinande, L.; Claes, K.B.; Van der Meulen, J.; Geboes, K.P. Long term response on Regorafenib in non-V600E BRAF mutated colon cancer: A case report. *BMC Cancer* **2019**, *19*, 1–5. [CrossRef] [PubMed]
14. Yoshino, K.; Manaka, D.; Kudo, R.; Kanai, S.; Mitsuoka, E.; Kanto, S.; Hamasu, S.; Konishi, S.; Nishitai, R. Metastatic colorectal cancer responsive to regorafenib for 2 years: A case report. *J. Med. Case Rep.* **2017**, *11*, 1–5. [CrossRef]
15. Roberto, M.; Falcone, R.; Mazzuca, F.; Archibugi, L.; Castaldi, N.; Botticelli, A.; Osti, M.F.; Marchetti, P. The role of stereotactic body radiation therapy in oligometastatic colorectal cancer: Clinical case report of a long-responder patient treated with regorafenib beyond progression. *Medicine* **2017**, *96*, e9023. [CrossRef] [PubMed]
16. Martinelli, E.; Sforza, V.; Cardone, C.; Capasso, A.; Nappi, A.; Martini, G.; Napolitano, S.; Rachiglio, A.M.; Normanno, N.; Cappabianca, S. Clinical outcome and molecular characterisation of chemorefractory metastatic colorectal cancer patients with long-term efficacy of regorafenib treatment. *ESMO Open* **2017**, *2*, e000177. [CrossRef] [PubMed]
17. Huang, D.; Sun, W.; Zhou, Y.; Li, P.; Chen, F.; Chen, H.; Xia, D.; Xu, E.; Lai, M.; Wu, Y. Mutations of key driver genes in colorectal cancer progression and metastasis. *Cancer Metastasis Rev.* **2018**, *37*, 173–187. [CrossRef]
18. Derakhshani, A.; Silvestris, N.; Hemmat, N.; Asadzadeh, Z.; Abdoli Shadbad, M.; Nourbakhsh, N.S.; Mobasheri, L.; Vahedi, P.; Shahmirzaie, M.; Brunetti, O. Targeting TGF-β-Mediated SMAD Signaling pathway via novel recombinant cytotoxin II: A potent protein from naja naja oxiana venom in Melanoma. *Molecules* **2020**, *25*, 5148. [CrossRef] [PubMed]
19. Zhang, D.; Qiu, X.; Li, J.; Zheng, S.; Li, L.; Zhao, H. TGF-β secreted by tumor-associated macrophages promotes proliferation and invasion of colorectal cancer via miR-34a-VEGF axis. *Cell Cycle* **2018**, *17*, 2766–2778. [CrossRef]
20. Song, Y.; Fu, Y.; Xie, Q.; Zhu, B.; Wang, J.; Zhang, B. Anti-angiogenic agents in combination with immune checkpoint inhibitors: A promising strategy for cancer treatment. *Front. Immunol.* **2020**, *11*, 1956. [CrossRef] [PubMed]
21. Brunetti, O.; Calabrese, A.; Palermo, L.; Solimando, A.G.; Argentiero, A. Long-term survival of an advanced colorectal cancer patient treated with Regorafenib: Case report and literature review. *Clin. Case Rep.* **2019**, *7*, 2379–2383. [CrossRef] [PubMed]
22. Silvestris, N.; Brunetti, O.; Pinto, R.; Petriella, D.; Argentiero, A.; Fucci, L.; Tommasi, S.; Danza, K.; De Summa, S. Immunological mutational signature in adenosquamous cancer of pancreas: An exploratory study of potentially therapeutic targets. *Expert Opin. Ther. Targets* **2018**, *22*, 453–461. [CrossRef] [PubMed]

23. Lai, Z.; Markovets, A.; Ahdesmaki, M.; Chapman, B.; Hofmann, O.; McEwen, R.; Johnson, J.; Dougherty, B.; Barrett, J.C.; Dry, J.R. VarDict: A novel and versatile variant caller for next-generation sequencing in cancer research. *Nucleic Acids Res.* **2016**, *44*, e108. [CrossRef] [PubMed]
24. Yang, H.; Wang, K. Genomic variant annotation and prioritization with ANNOVAR and wANNOVAR. *Nat. Protoc.* **2015**, *10*, 1556–1566. [CrossRef] [PubMed]
25. Gu, Z.; Eils, R.; Schlesner, M. Complex heatmaps reveal patterns and correlations in multidimensional genomic data. *Bioinform. (Oxf. Engl.)* **2016**, *32*, 2847–2849. [CrossRef]
26. Del Prete, M.; Giampieri, R.; Loupakis, F.; Prochilo, T.; Salvatore, L.; Faloppi, L.; Bianconi, M.; Bittoni, A.; Aprile, G.; Zaniboni, A. Prognostic clinical factors in pretreated colorectal cancer patients receiving regorafenib: Implications for clinical management. *Oncotarget* **2015**, *6*, 33982. [CrossRef]
27. Tabernero, J.; Lenz, H.-J.; Siena, S.; Sobrero, A.; Falcone, A.; Ychou, M.; Humblet, Y.; Bouché, O.; Mineur, L.; Barone, C. Analysis of circulating DNA and protein biomarkers to predict the clinical activity of regorafenib and assess prognosis in patients with metastatic colorectal cancer: A retrospective, exploratory analysis of the CORRECT trial. *Lancet Oncol.* **2015**, *16*, 937–948. [CrossRef]
28. Ricci, V.; Granetto, C.; Falletta, A.; Paccagnella, M.; Abbona, A.; Fea, E.; Fabozzi, T.; Nigro, C.L.; Merlano, M.C. Circulating cytokines and outcome in metastatic colorectal cancer patients treated with regorafenib. *World J. Gastrointest. Oncol.* **2020**, *12*, 301. [CrossRef]
29. Itatani, Y.; Kawada, K.; Sakai, Y. Transforming growth factor-β signaling pathway in colorectal cancer and its tumor microenvironment. *Int. J. Mol. Sci.* **2019**, *20*, 5822. [CrossRef]
30. Ullah, I.; Sun, W.; Tang, L.; Feng, J. Roles of Smads family and alternative splicing variants of Smad4 in different cancers. *J. Cancer* **2018**, *9*, 4018. [CrossRef]
31. Marisa, L.; de Reyniès, A.; Duval, A.; Selves, J.; Gaub, M.P.; Vescovo, L.; Etienne-Grimaldi, M.-C.; Schiappa, R.; Guenot, D.; Ayadi, M. Gene expression classification of colon cancer into molecular subtypes: Characterization, validation, and prognostic value. *PLoS Med.* **2013**, *10*, e1001453. [CrossRef]
32. Ahmed, S.; Bradshaw, A.-D.; Gera, S.; Dewan, M.Z.; Xu, R. The TGF-β/Smad4 signaling pathway in pancreatic carcinogenesis and its clinical significance. *J. Clin. Med.* **2017**, *6*, 5. [CrossRef] [PubMed]
33. Grothey, A.; Prager, G.; Yoshino, T. The mechanism of action of regorafenib in colorectal cancer: A guide for the community physician. *Clin. Adv. Hematol. Oncol. USA* **2019**, *17*, 1–19.

medicina MDPI

Article

Cell Models for Chromosome 20q11.21 Amplification and Drug Sensitivities in Colorectal Cancer

Ioannis A. Voutsadakis [1,2]

[1] Algoma District Cancer Program, Sault Area Hospital, 750 Great Northern Road, Sault Ste. Marie, ON P6B 0A8, Canada; ivoutsadakis@nosm.ca or ivoutsadakis@yahoo.com
[2] Section of Internal Medicine, Division of Clinical Sciences, Northern Ontario School of Medicine, Sudbury, ON P6B 0A8, Canada

Abstract: *Background and objectives*: The chromosome locus 20q11.21 is a commonly amplified locus in colorectal cancer, with a prevalence of 8% to 9%. Several candidate cancer-associated genes are transcribed from the locus. The therapeutic implications of the amplification in colorectal cancer remain unclear. *Materials and Methods*: Preclinical cell line models of colorectal cancer included in the Cancer Cell Line Encyclopedia (CCLE) collection were examined for the presence of amplifications in 20q11.21 genes. Correlations of the presence of 20q11.21 amplifications with gene essentialities and drug sensitivities were surveyed on salient databases for determination of therapeutic leads. *Results*: A significant subset of colorectal cancer cell lines in the CCLE (12 of 63 cell lines, 19%) bear amplifications of genes located at 20q11.21. Cancer-associated genes of the locus include *ASXL1*, *DNMT3B*, *BCL2L1*, *TPX2*, *KIF3B* and *POFUT1*. These genes are all amplified in the 12 cell lines, but they are variably over-expressed at the mRNA level, compared to non-amplified lines. 20q11.21 amplified cell lines are sensitive to various tyrosine kinase inhibitors and are resistant to chemotherapy drugs targeting the mitotic apparatus and microtubules. CRISPR and RNAi dependencies screening revealed, besides the β-catenin and *KRAS* genes, a few recurrent gene dependencies in more than one cell line, including *YAP1* and *JUP*. *Conclusions*: Cell line models of colorectal cancer with 20q11.21 gene amplifications display dependencies on the presence of specific genes and resistance or sensitivity to specific drugs and drug categories. Observations from in vitro models may form the basis for clinical drug development in this subtype of colorectal cancer. Genetic lesions conferring synthetic lethality to certain drugs or categories of drugs could be discovered with this approach.

Keywords: drug development; targeted therapies; preclinical candidates; databases; cancer therapeutics

Citation: Voutsadakis, I.A. Cell Models for Chromosome 20q11.21 Amplification and Drug Sensitivities in Colorectal Cancer. *Medicina* **2021**, 57, 860. https://doi.org/10.3390/medicina57090860

Academic Editors: Antonio M. Scanu, Maria Rosaria De Miglio and Konstantinos Dimas

Received: 10 June 2021
Accepted: 22 August 2021
Published: 24 August 2021

1. Introduction

Development of targeted therapies of cancer based on underlying molecular defects that drive carcinogenesis has improved cancer patient outcomes in recent years. New candidate drugs are often first tested in preclinical in vitro models consisting of cell lines cultured in highly artificial conditions [1]. Thus, their relevance for capturing the in vivo environment where the drugs under development will be acting is debatable. Moreover, tumors in situ consist not only of the tumor cells but also of a variety of supporting cells, as well as tumor attacking or tolerogenic immune cells and soluble factors (cytokines, chemokines and hormones) that both promote and antagonize tumor cell survival and proliferation [2]. Initial anti-cancer drug testing for identification of candidates for further development is performed with the aid of high-throughput screens, which, by their design, do not take into consideration the molecular characteristics of the tested cell lines. However, many of the tested drugs have a known molecular target, and their efficacy could be enhanced if the target is expressed and critical for the survival of neoplastic cells. In contrast, these drugs may be ineffective, even if pharmacologically and pharmacodynamically appropriate concentrations have been achieved, when the target is not expressed or is not critical in a given cancer [3]. Taking these factors into consideration, a molecularly informed drug development model starting early from the pre-clinical, in vitro phase, could aid in further clinical advancement of drugs with a drug–target paired model of development. A matched approach in the development process from early on could reduce the high attrition rates that hustle cancer drug discovery [4].

Amplification of specific loci is common in cancers and is often observed with tumor specificity, being more frequent in specific types of cancers. Perhaps one of the best-known amplifications with clinical therapeutic implications is observed in a subset of breast cancers at chromosome 17q. The locus includes oncogene *ERBB2*, encoding for HER2 receptor protein and leading to sensitivity to monoclonal antibodies and small-molecule tyrosine kinase inhibitors blocking the receptor [5]. In another cancer with high prevalence, colorectal adenocarcinoma, a commonly amplified region is at chromosome locus 20q11.21, which contains several potential oncogenic drivers and is amplified in about 10% of cases in the colorectal cancer cohort from TCGA [6]. In other cancers, amplifications of the 20q11.21 locus are rare and are encountered in 2.5% of head and neck carcinomas and in 1% to 1.5% of bladder carcinomas, small-cell lung carcinomas, esophagogastric and hepatobiliary cancers, which are the cancers with the higher prevalence of the amplification [7]. The amplified 20q11 region in colorectal cancers extends in many cases to neighboring loci, while in other cases, it is more restricted [7]. Genes that are located in the commonly amplified region include *ASXL1*, *DNMT3B*, *BCL2L1*, *TPX2*, *KIF3B* and *POFUT1*. The expression of resulting mRNA in 20q11.21 amplified colorectal cancers is variable. Some of the amplified genes are rarely over-expressed at the mRNA level, while others, such as *ASXL1*, *KIF3B* and *POFUT1*, are over-expressed in most amplified cancers. *ASXL1* is over-expressed in 88.9% of 20q11.21 amplified colorectal cancers in the TCGA cohort, while *POFUT1* is over-expressed in 90.5% of 20q11.21 amplified cases in the same cohort [7]. Genes that are over-expressed when amplified are putative drivers in the oncogenic process and may promote selection of the amplicon. This investigation examines drug sensitivity of colon and other cancer cell lines with amplification of genes at 20q11.21 for drugs that target driver genes of the amplicon. Dependencies of these cell lines to genomic alterations are also examined.

2. Methods

Cancer cell lines included in the current investigation constitute part of the Cancer Cell Line Encyclopedia (CCLE) collection [8]. The cBioportal Genomics Portal platform was used to probe CCLE for molecular abnormalities in colorectal cancer cell lines with amplification of genes located at 20q11.21 [9]. cBioCancer (http://www.cbioportal.org, accessed on 3 April 2021) is a user-friendly, open-access platform for genomic analysis of tumors and cancer cell lines [9]. Additionally, genomic data of colorectal cancer patients from The Cancer Genome Atlas (TCGA) study cohort [6] were analyzed using cBioportal. Subsets of cell lines and colorectal cancers with or without amplifications of genes at the 20q11.21 locus were identified. TCGA employs whole-exome sequencing to discover mutations, copy number alterations, and fusions in cohorts of patients with various types of cancer. Analysis of copy number alterations in TCGA is performed with the GISTIC (Genomic Identification of Significant Targets in Cancer) algorithm, in which a score of 2 or above denotes putative amplification of a gene [10]. An Aneuploidy Score (AS) is presented as a measure of chromosomal instability of cancers and is defined as the sum of the number of chromosome arms in each patient sample included in the study that display copy number alterations (gains or losses) [11]. A chromosome arm is considered copy number-altered based on the length of alterations, as calculated by the ABSOLUTE algorithm from Affymetrix 6.0 SNP arrays [12]. The definition of a somatic copy number alteration was set at more than 80% of the length of the arm. Alterations in 20% to 80% of a given arm length were considered inadequate to call, while chromosomal arms with somatic copy number alterations in less than 20% of the arm length were considered not altered. TCGA also includes mRNA expression analysis. The RSEM algorithm is used for normalization of mRNA expression [13].

The OncoKB knowledgebase is a database of cancer-related genes and classifies cancer-related genes as oncogenes or tumor suppressors [14]. OncoKB was scanned for any genes from the 20q11.21 locus included in the database and information was used to guide further analyses of putative drug dependencies.

The Genomics of Drug Sensitivity in Cancer (GDSC) dataset (www.cancerrxgene.org, accessed on 6 April 2021) was queried to obtain data on drug sensitivity of cell lines from colorectal cancer and other cancers with the 20q11.21 amplification [15]. Dependencies of these cell lines on specific genes was obtained from the Depmap portal that contains data from CRISPR arrays and RNA-interference arrays for CCLE cell lines [16,17]. These arrays screen cell lines for essential genes that are important for their survival and, as a result, their knock-down has a significant effect in their survival and proliferation in vitro [18–20].

Statistical comparisons of categorical data were carried out using Fisher's exact test or the χ^2 test. The Mann–Whitney U test was used to compare median values. All statistical comparisons were considered significant if $p < 0.05$.

3. Results

3.1. Cell Lines with 20q11.21 and Drug Sensitivity/Resistance In Vitro

Twelve cell lines among sixty-three colorectal cancer cell lines (19%) included in CCLE have amplifications of genes at the 20q11.21 locus, as assessed in cBioportal. All three genes from this locus that are listed as cancer-related at OncoKB knowledgebase (*ASXL1*, DNMT3B, BCL2L1) are amplified in the 12 cell lines (Table 1). Other genes of the locus with potential pathogenic importance in colorectal cancers, such as TPX2, *KIF3B* and *POFUT1*, are also amplified in the 12 cell lines. Another type of cancer with several cell lines displaying amplification of 20q11.21 genes is gastroesophageal carcinomas, where 14% to 15% of cell lines in CCLE carry amplifications of genes in the locus (Table 2). Interestingly, in contrast to colorectal cancer, gastroesophageal adenocarcinomas display 20q11.21 gene amplifications only in about 2.5% of clinical patient samples in TCGA gastroesophageal and gastric adenocarcinoma cohorts [21]. Similarly, non-small cell lung cancer (NSCLC) cell lines display a 15% to 18% prevalence of amplifications in 20q11.21 genes, while the prevalence of such amplifications in patients with either adenocarcinomas or squamous lung carcinomas is significantly lower [22,23]. The 20q11.21 locus amplification constitutes a recurrent copy number alteration confirmed in a pan-cancer cell line analysis performed in the GDSC database (feature cnaPANCAN363, www.cancerrxgene.org, accessed on 6 April 2021). Cancer cell lines with this recurrent copy number alteration, independently of primary type, display resistance to several currently used chemotherapy drugs, including the microtubule poisons vincristine, vinblastine and docetaxel, the topoisomerase II inhibitors teniposide and epirubicin and the DNA poisons temozolamide and actinomycin D (Table 3). In addition to microtubule inhibitor chemotherapeutics, several targeted mitotic inhibitors, including the Aurora A kinase inhibitor Alisertib, the kinesin protein family member 11 inhibitor Eg5 9814 and the CDC42BPA (CDC42 binding protein kinase alpha, also known as MRCK) inhibitor BDP-00009066, are associated with resistance in cell lines with 20q11.21 amplifications. Two inhibitors of DOT1L, an H3 histone methyltransferase with specificity for lysine 79 (H3K79), EPZ004777 and EPZ5676, also display resistance in these cell lines. A specific analysis of colorectal cancer cell lines with the 20q11.21 amplification shows that these cell lines are more resistant to several of those drugs (higher median IC_{50}) than colorectal cancer cell lines without the amplification, although, due to the smaller size of the cohort, differences are not statistically significant except for Alisertib and EPZ004777 (Table 3).

Table 1. Cell lines with 20q11.21 amplifications in CCLE. Twelve cell lines have amplifications of 20q11.21 genes in cBioportal. Information for the cell line characteristics is from Cell Model Passports. LN: Lymph node, MSI: Microsatellite instability, MSS: Microsatellite stable.

Name	DepMap ID	Provenance	Ploidy	Mutations/Mb	Driver Mutations	MSI Status
C2BBe1	ACH-000009	Colon primary	3.5	8.4	APC, TP53, SMAD4, CTNNB1	MSS
COLO678	ACH-000350	LN Metastasis	2.3	7.1	APC, KRAS	MSS
HT55	ACH-000926	Colon	3.4	33.7	APC, TP53, BRAF, FAT2, DNMT3A	MSS
LS1034	ACH-000252	Cecum	3.05	3.7	APC, TP53, KRAS, TCF7L2	MSS
LS123	ACH-000501	Colon	2.7	8.03	APC, TP53, KRAS, SMAD4, CTNND1	MSS
NCI-H508	ACH-000360	Abdominal wall metastasis	4.36	8.36	TP53, PIK3CA, CREBBP, RASA2	MSS
NCI-H747	ACH-000403	Cecum, LN metastasis	2.97	6.6	APC, TP53, KRAS	MSS
RCM1	ACH-000565	Rectum	3.08	9.73	TP53, APC, KRAS, FBXW7, SMAD4, FAT2, RBM10	MSS
SNU283	ACH-000708	Rectal, omental metastasis	3.5	10.9	ATM	MSS
SW1417	ACH-000236	Duke C, grade 3, primary tumor	3	7.06	APC, BRAF, TET2, EIF4G1	MSS
SW403	ACH-000820	Colon	3	NA	NA	NA
SW837	ACH-000421	Rectal	1.8	7	TP53, APC, KRAS, FBXW7, AMER1, ACVR2A	MSS

Table 2. Number of cell lines from different types of cancers with amplifications of genes located at chromosome 20q11.21. In parentheses are percentages of the total number of cell lines from each type of cancer included in CCLE.

Gene	Total	NSCLC	Esophagogastric	Colorectal	Melanoma	Pancreatic	Breast	Ovarian
ASXL1	96 (9.3)	15 (15.6)	14 (14.6)	12 (12.5)	7 (7.3)	7 (7.3)	6 (6.3)	6 (6.3)
BCL2L1	109 (10.6)	20 (18.3)	17 (15.6)	12 (11)	7 (6.4)	7 (6.4)	6 (5.5)	5 (4.6)
DNMT3B	86 (8.3)	13 (15.1)	12 (14)	12 (14)	7 (8.1)	6 (7)	5 (5.8)	3 (3.5)
TPX2	108 (10.5)	19 (17.6)	17 (15.7)	12 (11.1)	7 (6.5)	7 (6.5)	7 (6.5)	5 (4.6)
KIF3B	97 (9.4)	15 (15.5)	14 (14.4)	12 (12.4)	8 (8.2)	7 (7.2)	6 (6.2)	6 (6.2)
POFUT1	98 (9.5)	15 (15.3)	14 (14.3)	12 (12.2)	8 (8.2)	7 (7.1)	6 (6.1)	6 (6.1)

In contrast to promoting drug resistance, the cnaPANCAN363 copy number alteration (20q11.21 amplification) does not confer sensitivity to any of the 185 drugs tested in the Genomics of Drug Sensitivity in Cancer database (GDSC). However, individual colorectal cancer cell lines with the amplification display sensitivities to tested drugs (Table 4). Drug categories that show sensitivity in more than one cell line with the cnaPANCAN363 feature include receptor tyrosine kinase inhibitors, inhibitors of intracellular kinases (PI3K, mTOR, MEK and PKC) and lipid metabolism enzyme inhibitors of sphingosine kinase and Stearoyl-CoA desaturase. No specific drugs display sensitivity in more than three amplified colorectal cancer cell lines, suggesting that the mechanism of sensitivity may not be related to the 20q11.21 amplification, that they all possess.

Table 3. Drug resistance associated with amplifications of 20q11.21 (feature cnaPANCAN363) in the whole CCLE cohort of cell lines, independently of primary type. In parentheses are respective values for the colorectal cancer cell lines cohort. Bold denotes significant in both the pan-cancer and the colorectal cancer cohort. IC_{50}: Inhibitory Concentration 50%.

Drug	Mechanism of Action	Group with cnaPANCAN363 (Median IC_{50} in μmol)	Group without cnaPANCAN363 (Median IC_{50} in μmol)	*p*-Value
Vinblastine	Microtubule poison	0.072 (0.27)	0.016 (0.019)	4×10^{-7} (0.17)
Vincristine	Microtubule poison	0.78 (0.93)	0.1 (0.24)	0.00001 (0.6)
Docetaxel	Microtubule poison	0.021 (0.05)	0.0009 (0.01)	6×10^{-6} (0.09)
Teniposide	Podophylotoxin, Topoisomerase II inhibitor	4.25 (9.4)	1.41 (4.4)	3×10^{-6} (0.15)
Epirubicin	Topoisomerase II inhibitor	0.65 (2.2)	0.31 (0.47)	9×10^{-7} (0.13)
Dactinomycin	DNA intercalator	0.017 (0.018)	0.007 (0.009)	0.00002 (0.6)
Temozolamide	Alkylating agent	616 (1181)	370 (626)	2.9×10^{-7} (0.11)
Fulvestrant	SERD	30.88 (51.73)	17.38 (26.12)	9.4×10^{-8} (0.16)
Nutlin-3a	Apoptosis sensitizer	275.3 (442.9)	112.9 (197.1)	5.9×10^{-8} (0.06)
Alisertib	AURKA inhibitor	18.5 (63.5)	5.9 (6.7)	**3×10^{-6} (0.01)**
EPZ004777	DOT1L inhibitor	270.2 (377.8)	162.3 (249.8)	**1.2×10^{-7} (0.03)**
EPZ5676	DOT1L inhibitor	461.9 (545)	243.3 (331)	5.7×10^{-7} (0.22)
YK-4-279	ETS inhibitor	22.1 (62.6)	7.7 (5.8)	3.6×10^{-7} (0.056)
Gallibiscoquinazole	?	24.5 (27.6)	12.1 (12.9)	3.6×10^{-7} (0.22)
Pevonedistat	NEDD8 activating enzyme inhibitor	7.74 (8.1)	1.61 (3.11)	9.7×10^{-7} (0.56)
BDP-00009066	MRCK inhibitor	15.97 (12.1)	9.71 (10.6)	2×10^{-5} (0.36)
Eg5 9814	KSP11 kinesin inhibitor	0.1 (0.57)	0.03 (0.05)	4.6×10^{-7} (0.11)
AGI-5198	IDH inhibitor	177.3 (184.9)	94.3 (114.4)	6×10^{-8} (0.09)

Table 4. Top-5 drugs with z-scores in cell lines with 20q11.21 amplifications. Data are from Genomics of Drug Sensitivity in Cancer datasets (GDSC1 and GDSC2, www.cancerrxgene.org, accessed on 6 April 2021).

Cell Line	Drug Name	Drug Targets	IC_{50} (μM)	Z Score	Dataset
C2BBe1	Selumetinib	MEK1, MEK2	0.19	−2.78	GDSC1
	AZD8186	PI3Kalpha, PI3Kbeta	0.90	−2.72	GDSC2
	SB505124	TGFBR1, ACVR1B, ACVR1C	0.43	−2.63	GDSC2
	TGX221	PI3Kbeta	4.99	−2.09	GDSC1
	Vismodegib	SMO	13.99	−2.03	GDSC1
COLO678	Selumetinib	MEK1, MEK2	0.91	−1.73	GDSC1
	Cetuximab	EGFR	121.42	−1.25	GDSC1
	Afatinib	ERBB2, EGFR	0.45	−1.14	GDSC1
	Selumetinib	MEK1, MEK2	0.71	−1.02	GDSC1
	Dyrk1b_0191	DYRK1B	3.32	−0.96	GDSC1
HT55	Bryostatin 1	PKC	0.01	−2.58	GDSC1
	Linsitinib	IGF1R	1.67	−2.52	GDSC2
	BMS-754807	IGF1R, IR	0.02	−2.27	GDSC2
	CAY10566	Stearoyl-CoA desaturase	1.22	−2.02	GDSC1
	Picolinic-acid	Inflammatory related	38.76	−1.88	GDSC2

Table 4. *Cont.*

Cell Line	Drug Name	Drug Targets	IC_{50} (μM)	Z Score	Dataset
LS1034	Linsitinib	IGF1R	0.24	−2.46	GDSC1
	CAY10566	Stearoyl-CoA desaturase	1.02	−2.16	GDSC1
	BMS-754807	IGF1R, IR	0.06	−2.06	GDSC1
	VX-11e	ERK2	1.10	−1.86	GDSC1
	STF-62247	Autophagy inducer	18.62	−1.52	GDSC1
LS123	Acetalax		0.37	−2.68	GDSC2
	TANK_1366	Tankyrase 1/2 (PARP5a, PARP5b)	1.77	−2.13	GDSC1
	PFI-3	SMARCA2, SMARCA4, PB1	26.60	−2.00	GDSC1
	QL-VIII-58	MTOR, ATR	0.04	−1.44	GDSC1
	Sepantronium bromide	BIRC5	0.00	−1.41	GDSC2
NCI-H508	Afatinib	ERBB2, EGFR	0.04	−2.81	GDSC1
	Afatinib	ERBB2, EGFR	0.07	−2.71	GDSC2
	Gefitinib	EGFR	0.23	−2.12	GDSC1
	Pictilisib	PI3K (class 1)	0.18	−2.00	GDSC1
	MK-2206	AKT1, AKT2	0.87	−1.97	GDSC2
NCI-H747	Amuvatinib	KIT, PDGFRA, FLT3	0.93	−2.41	GDSC1
	Enzastaurin	PKCB	0.86	−2.30	GDSC1
	Selumetinib	MEK1, MEK2	0.41	−2.28	GDSC1
	Refametinib	MEK1, MEK2	0.10	−2.18	GDSC1
	Selumetinib	MEK1, MEK2	0.12	−1.94	GDSC1
RCM1	Selumetinib	MEK1, MEK2	0.46	−2.19	GDSC1
	Refametinib	MEK1, MEK2	0.18	−1.83	GDSC1
	CCT007093	PPM1D	15.57	−1.80	GDSC1
	AS605240	PI3Kgamma	1.60	−1.69	GDSC1
	BPD-00008900		16.98	−1.61	GDSC2
SNU283	NA				
SW1417	WEHI-539	BCL-XL	0.33	−2.48	GDSC2
	Sphingosine Kinase 1 Inhibitor II	Sphingosine Kinase	10.20	−2.05	GDSC1
	CHIR-99021	GSK3A, GSK3B	3.07	−1.99	GDSC1
	Navitoclax	BCL2, BCL-XL, BCL-W	0.28	−1.61	GDSC2
	SN-38	TOP1	0.00	−1.43	GDSC1
SW403	NA				
SW837	AS605240	PI3Kgamma	1.17	−1.90	GDSC1
	Dihydrorotenone		0.23	−1.49	GDSC2
	Acetalax		7.31	−1.25	GDSC2
	VX-11e	ERK2	3.49	−1.06	GDSC1
	Trametinib	MEK1, MEK2	0.07	−0.98	GDSC2

3.2. Increased mRNA Expression of Genes from 20q11.21 and Targeted Drugs

mRNA expression of genes at 20q11.21 in 12 colorectal cancer cell lines with 20q11.21 amplification was compared with the corresponding expression in 12 randomly selected colorectal cancer cell lines from CCLE without amplification in the locus. Among the six genes located at 20q11.21 with potential cancer pathogenesis interest, BCL2L1, *POFUT1* and *KIF3B* were over-expressed at the mRNA level in 20q11.21 amplified cell lines compared with non-amplified cell lines (Table 5). Over-expression of amplified genes is partially overlapping with the over-expression of the genes in 20q11.21 amplified clinical samples of colorectal cancer patients, where *POFUT1* but also *ASXL1*, and, in fewer cases, *KIF3B* and TPX2, are often over-expressed (Table 5).

Table 5. Comparison of mRNA expression of genes at 20q11.21 in cell lines with and without 20q11.21 amplifications.

	ASXL1	*BCL2L1*	*DNMT3B*	*TPX2*	*POFUT1*	*KIF3B*
Median cell lines with 20q11.21 amplification	19.51098	114.1379	2.201885	102.2259	32.98115	24.43462
Median cell lines without 20q11.21 amplification	16.05503	52.77	3.028	78.56	18.96	17.97
U	53	31	65	43	28	31
p (critical value)	>0.05 (37)	<0.05 (37)	>0.05 (37)	>0.05 (37)	<0.05 (37)	<0.05 (37)
Z (p)	1.06 (0.28)	2.3 (0.01)	−0.3 (0.7)	1.6 (0.09)	2.5 (0.01)	2.33 (0.01)
Over-expression in patients (%)	88.9	43.4	58.5	69.8	90.5	79.2

Based on the mRNA over-expressions of genes at 20q11.21, the sensitivity of cell lines to BCL-xL inhibitors, Notch inhibitors (Notch pathway is activated by *POFUT1* enzyme) and mitotic spindle inhibitors was evaluated at the GDSC. A notable drug in these categories displaying sensitivity in amplified cell lines compared to non-amplified cell lines is the BCL2 family inhibitor WEHI-539 (median IC_{50} in amplified lines 11.6 μmol versus 42.5 μmol in non-amplified cell lines, $p = 0.04$). In contrast, other BCL2 family inhibitors examined such as venetoclax and navitoclax showed no sensitivity in amplified cell lines or even a trend for resistance compared to non-amplified lines. Colorectal cancer cell lines with the 20q11.21 amplification displayed resistance to the microtubule polymerization stabilizer epothilone B, compared with non-amplified cell lines (median IC_{50} in amplified lines 0.028 μmol versus 0.003 μmol in non-amplified cell lines, $p = 0.01$). Z-LLNie-CHO, a γ secretase inhibitor of the Notch cascade, displayed a non-significant trend towards resistance in amplified cell lines (median IC_{50} in amplified lines 7.36 μmol versus 1.96 μmol in non-amplified cell lines, $p = 0.23$).

3.3. CRISPR Microarray Dependencies of 20q11.21 Amplified Colorectal Cancer Cell Lines

An evaluation of the CRISPR preferentially essential genes and RNA-interference screening of colorectal cancer cell lines with the 20q11.21 amplification disclosed a few recurrent genes in more than one line, that include *CTNNB1*, encoding for β-catenin, the WNT pathway transcription factor *TCF7L2* and oncogene *KRAS* (Table 6). In addition, BCL2L1, the gene JUP, encoding for Junction Plakoglobin (also called γ-catenin), and YAP1 (Yes-associated protein), encoding for a Hippo pathway nuclear regulator, are among additional recurrent dependency genes in 20q11.21 amplified cell lines. A similar CRISPR KO screen from project SCORE that excluded known core fitness genes of colorectal cancer, disclosed a few non-overlapping recurrent genes in colorectal cancer cell lines with 20q11.21 amplification, including DONSON (downstream of the SON gene, DNA replication fork stabilization factor), SNAP23 (synaptosome-associated protein 23) and HMGCS1 (3-hydroxy-3-methylglutaryl-CoA synthase 1).

Table 6. CRISPR and RNAi dependencies of colorectal cancer cell lines with 20q11.21 amplification. CRISPR data are from CRISPR (Avana) Public 21Q1 screen and RNAi data are from a combined dataset from Board Institute, Novartis and Marcotte et al. [20], as compiled in the Cell Model Passports site (cellmodelpassports.sanger.ac.uk, accessed on 6 April 2021). NA: Not available. Recurrent genes are in bold.

Cell Line	Top-10 CRISPR Preferentially Essential Genes	RNAi Screen
C2BBe1	**JUP**, NXT1, PPIL1, KDSR, IER3IP1, SCAP, BCAS2, ERBB2, ARF4, EGFR	**CTNNB1**, **JUP**, BCAS2, URI1, STK32A, TACR3, DOT1L, SRPK1, CMAS, MED6
COLO678	**CTNNB1**, **TCF7L2**, FAM50A, PSMB6, RPL37, CCDC86, NUP43, OGDH, YPEL5, MMS22L	NXF1, **CTNNB1**, CDC40, SNRPF, VPS28, **KRAS**, DDB1, SF3A1, TSG101, RANBP2
HT55	**CTNNB1**, IRS2, **TCF7L2**, TUBB4B, ERMP1, FASN, NXT1, ACTB, IQGAP1, STXBP3	**CTNNB1**, CAPZB, ABCE1, MAGOHB, FASN, DARS1, CFL1, SRSF3, VARS1, NXT1
LS1034	NA	NA
LS123	NA	NA
NCI-H508	MYBL2, SLC22A20P, TYMS, ANKRD20A19P, SKP1, PSMA3, **CTNNB1**, **YAP1**, EFCAB8, EGFR	NA
NCI-H747	FERMT1, VPS4A, VPS4B, **KRAS**, RAB6A, NONO, SCAP, SNRPB2, **YAP1**, RHOA	**YAP1**, **KRAS**, **JUP**, UBC, **BCL2L1**, BUB1B, FERMT1, ATP6V1B2, GPX4, UBA1
RCM1	NA	NA
SNU283	NA	NA
SW1417	CDC40, PPIE, KHSRP, EFCAB8, PPP2CA, PPWD1, EIF4A3, MED11, OR56B1, CAPZB	NA
SW403	MOCS3, GRB14, LONP1, QRFPR, FKBP1A, C2orf50, ADTRP, GNB1L, SNAP91, BOC	**KRAS**, **CTNNB1**, GPX4, USP5, WDR18, **BCL2L1**, UFD1, KDM2A, TSPAN7, FASN
SW837	NF2, INTS6, SMARCA4, **KRAS**, **JUP**, ATP7A, MED16, CAB39, DBF4, TSC2	MED1, RBM19, GTF3A, CRNKL1, SMARCA4, **KRAS**, **YAP1**, TRERF1, CDC7, SLBP

4. Discussion

Amplifications of loci at the long arm of chromosome 20 are common in colorectal cancers. Genes at locus 20q11.21 are listed among the most commonly amplified genes in colorectal cancers. Colorectal cancers with the 20q11.21 amplification present at a similar stage with 20q11.21 non-amplified colorectal cancers and have a similar overall survival [7]. However, when metastatic, colorectal cancers with 20q11.21 amplification have a better survival compared with non-amplified metastatic counterparts. Moreover, 20q11.21 amplified colorectal cancer rarely harbor mutations in DNA damage response and mismatch repair-related genes compared with 20q11.21 non-amplified colorectal cancers, and have a lower tumor mutation burden [7]. The 20q11.21 locus harbors several cancer-associated genes with potential oncogenic properties. These include the epigenetic regulators *ASXL1* and DNMT3B, the apoptosis regulator BCL2L1, the microtubule and mitotic spindle-associated proteins TPX2 and *KIF3B* and the enzyme fucosyl-transferase *POFUT1*. Proteins encoded by cancer-associated genes at 20q11.21 are expressed in most cases of colorectal cancers in the Human Protein Atlas [24]. Variability in the intensity of staining is observed that may underline differences in translation, in addition to gene dosage. Amplification of one or more of these genes, as a consequence of 20q11.21 locus amplification, may be the event that favors selection of the amplification and results in its comparatively high prevalence in colorectal cancer and colorectal cancer cell lines. In addition, such a driver event could be a therapeutic target for the subset of cancers carrying the amplification. Targeting a driver defect in the specific subsets of cancers that bear it would be effective in these patients and avoid treatment toxicity in the rest of the patients with no amplification of the locus. Moreover, it would help the development of targeted drugs, as a therapeutic benefit would be difficult to discern if the target population in trials is diluted by non-responders. With these considerations, the current investigation sought to take advantage of databases of in vitro cancer models in an attempt to provide a framework of sensitivities for colorectal cancers with the 20q11.21 amplification. The main findings of the current study are manifold. First, it was shown that a subset of colorectal cancer cell lines bear the 20q11.21, capturing the corresponding colorectal cancer biology of a subset of patients. Second, as a group, cancer cell lines with 20q11.21 amplifications tend to be more resistant to microtubule inhibitors, topoisomerase II inhibitors and some DNA alkylators. In addition, resistance to targeted Aurora A kinase inhibitors and kinesin inhibitors is observed in these cell lines. In contrast, the amplification does not endow 20q11.21 amplified colorectal cancer cell lines with sensitivity to any of the drugs checked in the database. However, individual cell lines with the amplification display sensitivity to assayed drugs, including kinase inhibitors and lipid metabolism inhibitors, albeit only in a few cell lines in each case. Recurrent dependencies of cell lines with the amplification include the genes for YAP1, JUP (γ-catenin), BCL2L1, DONSON, SNAP23 and HMGCS1. These dependencies may provide clues for additional therapeutic interventions.

Amplification of 20q11.21 is observed in a significant minority of cell lines beyond colorectal cancer, such as esophagogastric and non-small lung cancers, despite the low prevalence of the amplification in patient samples from these cancers. This may suggest that the amplified segment genes confer advantage in these cancers in vitro, which is not essential in vivo. In contrast, in colorectal cancers, 20q11.21 amplifications are advantageous both in vitro and in vivo. Drug sensitivities and resistance mechanisms stemming from over-expression of genes amplified from the locus would be expected to present independently of cell line origin. With this rationale, and in order to increase statistical power from an increased number of cell lines, this study compared all 20q11.21 amplified cell lines from CCLE included in the GDSC project to those without the 20q11.21 amplification. Comparisons were, then, focused on the colorectal cell lines subsets. Results were concordant in the two comparisons, although, as expected, due to smaller numbers, they were mostly not statistically significant in the latter set of comparisons.

A different approach that could help with a better targeting of therapies to appropriate subsets of patients is by categorizing colorectal cancers to genomic subsets. Colorectal cancers have been categorized according to genomic profiles into four consensus molecular subtypes (CMS1 to 4) [25]. Cancers with 20q11.21 amplifications represent a subset of the most common canonical CSM2 cancers [7]. Characteristics of the CMS2 group include left colon laterality in 77% of cases, high level of chromosomal instability, high frequency of APC mutations leading to WNT pathway activation and lower levels of MSI lesions. KRAS mutations, BRAF mutations and SMAD4 mutations are less frequent in CSM2 colorectal cancers than in other subtypes [7,26]. Thus, CMS2 cancers and, among them, cancers with 20q11.21, would be expected to respond to EGFR-targeting therapies. Data from drug sensitivity analysis of several 20q11.21 amplified cell lines concur with this assumption of sensitivity to EGFR and downstream kinase inhibitors (Table 4). However, the molecular consensus classification does not provide any additional guidance for currently available therapies, and the need for new options based on biomarkers of efficacy remains. It is reassuring that none of the chemotherapy drugs identified as being associated with resistance in 20q11.21 cancers are used clinically in colorectal cancer. The therapeutic implications of resistance to targeted mitotic inhibitors, including Aurora A kinase and kinesin inhibitors, is of interest and suggests that TPX2 and *KIF3B* amplifications may be involved in dysregulation of mitosis, leading to the observed mitosis inhibitors' resistance.

Dependencies of 20q11.21 amplified cell lines on particular genes and their products as derived from CRISPR and RNAi arrays could inform development of therapies based on synthetic lethalities. Yes-associated protein 1 (YAP1), a transcription factor of the Hippo pathway, comes up in the dependency screening of 20q amplified colorectal cancer cell lines, confirming a key role of the pathway in colorectal cancer. YAP1 co-operates with transcription factors TAZ and TEAD in transcription of genes involved in proliferation following tissue damage and promoting regeneration in the gut [27]. In colorectal cancer, aberrant signals from cancer-associated pathways, such as WNT and activated KRAS, activate Hippo to promote tumor growth and metastasis [28]. It is intriguing that the apoptosis inhibitor BCL2 is among the target genes of YAP1, a fact that could contribute to YAP1 dependency in 20q11.21 amplified cancers, given that the related BCL2L1 protein is dysregulated in these cancers [29]. Thus, interruption of Hippo signaling could impede, at least partially, aberrant cancer cell signals. CMS2 cancers are characterized by WNT pathway activation and thus a downstream activation of Hippo. Similarly, the JUP gene encoding for the β-catenin homolog, junction plakoglobin (also called γ-catenin), is also shown to be a dependence gene in a subset of 20q11.21 amplified colorectal cancer cell lines. γ-catenin has parallel roles with β-catenin in cell adhesion and WNT signaling [30]. In addition to adherens junctions, γ-catenin has a role in desmosomes [31]. These data suggest that the network of proteins associated with alternative fates of Wnt signaling is an important node in 20q11.21 amplified colorectal cancers and a candidate for therapeutic interventions.

DONSON, one of the discovered dependencies present in 2 of 7 tested amplified cell lines, is a gene of unknown function, playing a role in DNA replication and the stabilization and protection of stalled replication forks. Mutations of the gene are associated with the microcephaly-micromelia syndrome [32]. In cancer, DONSON could be helpful in preventing apoptosis during aberrant DNA replication. However, the mechanism through which 20q11.21 amplified cancers and cancers with increased chromosomal instability in general could be associated with DONSON dependence remains to be unveiled.

SNAP23, another dependency gene present in 2 of 7 tested amplified cell lines, encodes for one of the proteins of the cellular machinery for membrane fusion and exocytosis. It is also involved in cell signaling, promoting malignant cell motion, and through this mechanism, it may favor metastasis [33].

HMGCS1 is a mevalonate precursor enzyme and catalyzes the conversion of two molecules of acetoacetyl-CoA to form 3-hydroxy-3-methyl-glutaryl-CoA (HMG-CoA), the precursor of cholesterol production [34]. HMGCS1 plays a role in breast cancer stem cells, and its downregulation decreased the stem cell fraction of both luminal and basal breast cancer cells [35]. Cholesterol biosynthesis is targeted by statins, a class of cholesterol-lowering drugs, and attempts at repurposing these drugs for cancer are in progress [36]. Interestingly, drugs targeting other lipid metabolism enzymes show activity in some 20q11.21 amplified cell lines, suggesting lipid metabolism as a possible target in these cancers. Repurposing of drugs already used for other indications for well-defined subsets of colorectal cancers would present significant advantages from financial and patient safety perspectives.

Funding: This research received no external funding.

Institutional Review Board Statement: Not applicable.

Informed Consent Statement: Not applicable.

Data Availability Statement: No data additional data related to this report are available.

Conflicts of Interest: The author declares no conflict of interest.

References

1. Wilding, J.L.; Bodmer, W.F. Cancer cell lines for drug discovery and development. *Cancer Res.* **2014**, *74*, 2377–2384. [CrossRef]
2. Goto, T. Patient-Derived Tumor Xenograft Models: Toward the Establishment of Precision Cancer Medicine. *J. Personal. Med.* **2020**, *10*, 64. [CrossRef] [PubMed]
3. Ruiz-Moreno, A.J.; Torres-Barrera, P.; Velázquez-Paniagua, M.; Dömling, A.; Velasco-Velázquez, M.A. Guide for Selection of Relevant Cell Lines During the Evaluation of new Anti-Cancer Compounds. *Anticancer Agents Med. Chem.* **2018**, *18*, 1072–1081. [CrossRef] [PubMed]
4. Moreno, L.; Pearson, A.D. How can attrition rates be reduced in cancer drug discovery? *Expert Opin. Drug. Discov.* **2013**, *8*, 363–368. [CrossRef] [PubMed]
5. Voutsadakis, I.A. Chromosome 17 centromere amplification and chromosomal instability (CIN) in breast cancer: Pathogenic and therapeutic implications. *Neoplasma* **2019**, *66*, 859–869. [CrossRef]

6. Cancer Genome Atlas Network. Comprehensive molecular characterization of human colon and rectal cancer. *Nature* **2012**, *487*, 330–337. [CrossRef] [PubMed]
7. Voutsadakis, I.A. Chromosome 20q11.21 Amplifications in Colorectal Cancer. *Cancer Genom. Proteome* **2021**, *18* (Suppl. 3), 487–496. [CrossRef]
8. Barretina, J.; Caponigro, G.; Stransky, N.; Venkatesan, K.; Margolin, A.A.; Kim, S.; Wilson, C.J.; Lehár, J.; Kryukov, G.V.; Sonkin, D.; et al. The Cancer Cell Line Encyclopedia enables predictive modelling of anticancer drug sensitivity. *Nature* **2012**, *483*, 603–607. [CrossRef]
9. Gao, J.; Aksoy, B.A.; Dogrusoz, U.; Dresdner, G.; Gross, B.; Sumer, S.O.; Sun, Y.; Jacobsen, A.; Sinha, R.; Larsson, E.; et al. Integrative analysis of complex cancer genomics and clinical profiles using the cBioPortal. *Sci. Signal.* **2013**, *6*, pl1. [CrossRef]
10. Mermel, C.H.; Schumacher, S.E.; Hill, B.; Meyerson, M.L.; Beroukhim, R.; Getz, G. GISTIC2.0 facilitates sensitive and confident localization of the targets of focal somatic copy-number alteration in human cancers. *Genome Biol.* **2011**, *12*, R41. [CrossRef]
11. Taylor, A.M.; Shih, J.; Ha, G.; Gao, G.F.; Zhang, X.; Berger, A.C.; Schumacher, S.E.; Wang, C.; Hu, H.; Liu, J.; et al. Genomic and Functional Approaches to Understanding Cancer Aneuploidy. *Cancer Cell* **2018**, *33*, 676–689.e3. [CrossRef] [PubMed]
12. Carter, S.L.; Cibulskis, K.; Helman, E.; McKenna, A.; Shen, H.; Zack, T.; Laird, P.W.; Onofrio, R.C.; Winckler, W.; Weir, B.A.; et al. Absolute quantification of somatic DNA alterations in human cancer. *Nat. Biotechnol.* **2012**, *30*, 413–421. [CrossRef] [PubMed]
13. Li, B.; Dewey, C.N. RSEM: Accurate transcript quantification from RNA-seq data with or without a reference genome. *BMC Bioinform.* **2011**, *12*, 323. [CrossRef] [PubMed]
14. Chakravarty, D.; Gao, J.; Phillips, S.M.; Kundra, R.; Zhang, H.; Wang, J.; Rudolph, J.E.; Yaeger, R.; Soumerai, T.; Nissan, M.H.; et al. OncoKB: A Precision Oncology Knowledge Base. *JCO Precis. Oncol.* **2017**, *2017*. [CrossRef] [PubMed]
15. Iorio, F.; Knijnenburg, T.A.; Vis, D.J.; Bignell, G.R.; Menden, M.P.; Schubert, M.; Aben, N.; Gonçalves, E.; Barthorpe, S.; Lightfoot, H.; et al. A Landscape of Pharmacogenomic Interactions in Cancer. *Cell* **2016**, *166*, 740–754. [CrossRef] [PubMed]
16. Behan, F.M.; Iorio, F.; Picco, G.; Gonçalves, E.; Beaver, C.M.; Migliardi, G.; Santos, R.; Rao, Y.; Sassi, F.; Pinnelli, M.; et al. Prioritization of cancer therapeutic targets using CRISPR-Cas9 screens. *Nature* **2019**, *568*, 511–516. [CrossRef]
17. Boehm, J.S.; Garnett, M.J.; Adams, D.J.; Francies, H.E.; Golub, T.R.; Hahn, W.C.; Iorio, F.; McFarland, J.M.; Parts, L.; Vazquez, F. Cancer research needs a better map. *Nature* **2021**, *589*, 514–516. [CrossRef]
18. Van der Meer, D.; Barthorpe, S.; Yang, W.; Lightfoot, H.; Hall, C.; Gilbert, J.; Francies, H.E.; Garnett, M.J. Cell Model Passports-a hub for clinical, genetic and functional datasets of preclinical cancer models. *Nucleic Acids Res.* **2019**, *47*, D923–D929. [CrossRef]
19. Tsherniak, A.; Vazquez, F.; Montgomery, P.G.; Weir, B.A.; Kryukov, G.; Cowley, G.S.; Gill, S.; Harrington, W.F.; Pantel, S.; Krill-Burger, J.M.; et al. Defining a Cancer Dependency Map. *Cell* **2017**, *170*, 564–576.e16. [CrossRef]
20. Marcotte, R.; Sayad, A.; Brown, K.R.; Sanchez-Garcia, F.; Reimand, J.; Haider, M.; Virtanen, C.; Bradner, J.E.; Bader, G.D.; Mills, G.B.; et al. Functional Genomic Landscape of Human Breast Cancer Drivers, Vulnerabilities, and Resistance. *Cell* **2016**, *164*, 293–309. [CrossRef]
21. Cancer Genome Atlas Research Network. Comprehensive molecular characterization of gastric adenocarcinoma. *Nature* **2014**, *11*, 202–209. [CrossRef]
22. Cancer Genome Atlas Research Network. Comprehensive genomic characterization of squamous cell lung cancers. *Nature* **2012**, *489*, 519–525. [CrossRef]
23. Cancer Genome Atlas Research Network. Comprehensive molecular profiling of lung adenocarcinoma. *Nature* **2014**, *511*, 543–550. [CrossRef]
24. Uhlén, M.; Fagerberg, L.; Hallström, B.M.; Lindskog, C.; Oksvold, P.; Mardinoglu, A.; Sivertsson, Å.; Kampf, C.; Sjöstedt, E.; Asplund, A.; et al. Proteomics. Tissue-based map of the human proteome. *Science* **2015**, *347*, 1260419. [CrossRef]
25. Guinney, J.; Dienstmann, R.; Wang, X.; de Reyniès, A.; Schlicker, A.; Soneson, C.; Marisa, L.; Roepman, P.; Nyamundanda, G.; Angelino, P.; et al. The consensus molecular subtypes of colorectal cancer. *Nat. Med.* **2015**, *21*, 1350–1356. [CrossRef]
26. Lee, M.S.; Menter, D.G.; Kopetz, S. Right Versus Left Colon Cancer Biology: Integrating the Consensus Molecular Subtypes. *J. Natl. Compr. Cancer Netw.* **2017**, *15*, 411–419. [CrossRef]
27. Guillermin, O.; Angelis, N.; Sidor, C.M.; Ridgway, R.; Baulies, A.; Kucharska, A.; Antas, P.; Rose, M.R.; Cordero, J.; Sansom, O.; et al. Wnt and Src signals converge on YAP-TEAD to drive intestinal regeneration. *EMBO J.* **2021**, e105770. [CrossRef] [PubMed]
28. Thompson, B.J. YAP/TAZ: Drivers of Tumor Growth, Metastasis, and Resistance to Therapy. *Bioessays* **2020**, *42*, e1900162. [CrossRef] [PubMed]
29. Jin, L.; Chen, Y.; Cheng, D.; He, Z.; Shi, X.; Du, B.; Xi, X.; Gao, Y.; Guo, Y. YAP inhibits autophagy and promotes progression of colorectal cancer via upregulating Bcl-2 expression. *Cell Death Dis.* **2021**, *12*, 457. [CrossRef]
30. Nagel, J.M.; Lahm, H.; Ofner, A.; Göke, B.; Kolligs, F.T. γ-Catenin acts as a tumor suppressor through context-dependent mechanisms in colorectal cancer. *Int. J. Colorectal Dis.* **2017**, *32*, 1243–1251. [CrossRef]
31. Ben-Ze'ev, A.; Geiger, B. Differential molecular interactions of beta-catenin and plakoglobin in adhesion, signaling and cancer. *Curr. Opin. Cell Biol.* **1998**, *10*, 629–639. [CrossRef]
32. Karaca, E.; Posey, J.E.; Bostwick, B.; Liu, P.; Gezdirici, A.; Yesil, G.; Coban Akdemir, Z.; Bayram, Y.; Harms, F.L.; Meinecke, P.; et al. Biallelic and De Novo Variants in DONSON Reveal a Clinical Spectrum of Cell Cycle-opathies with Microcephaly, Dwarfism and Skeletal Abnormalities. *Am. J. Med. Genet. A* **2019**, *179*, 2056–2066. [CrossRef] [PubMed]
33. Williams, K.C.; Coppolino, M.G. SNARE-dependent interaction of Src, EGFR and β1 integrin regulates invadopodia formation and tumor cell invasion. *J. Cell Sci.* **2014**, *127 Pt 8*, 1712–1725. [CrossRef]

34. Buhaescu, I.; Izzedine, H. Mevalonate pathway: A review of clinical and therapeutical implications. *Clin. Biochem.* **2007**, *40*, 575–584. [CrossRef]
35. Walsh, C.A.; Akrap, N.; Garre, E.; Magnusson, Y.; Harrison, H.; Andersson, D.; Jonasson, E.; Rafnsdottir, S.; Choudhry, H.; Buffa, F.; et al. The mevalonate precursor enzyme HMGCS1 is a novel marker and key mediator of cancer stem cell enrichment in luminal and basal models of breast cancer. *PLoS ONE* **2020**, *15*, e0236187. [CrossRef] [PubMed]
36. Iannelli, F.; Lombardi, R.; Milone, M.R.; Pucci, B.; De Rienzo, S.; Budillon, A.; Bruzzese, F. Targeting Mevalonate Pathway in Cancer Treatment: Repurposing of Statins. *Recent Pat. Anticancer Drug Discov.* **2018**, *13*, 184–200. [CrossRef] [PubMed]

medicina MDPI

Article

An Insight into Deficient Mismatch Repair Colorectal Cancer Screening in a Romanian Population—A Bi-Institutional Pilot Study

Cristina Lungulescu [1,†], Vlad Mihai Croitoru [2,†], Simona Ruxandra Volovat [3], Irina Mihaela Cazacu [2], Adina Turcu-Stiolica [4], Dan Ionut Gheonea [5], Daniel Sur [6,7,*] and Cristian Virgil Lungulescu [8]

1 Doctoral School, University of Medicine and Pharmacy Craiova, 2 Petru Rares Str., 200349 Craiova, Romania; cristina.lungulescu@yahoo.com
2 Department of Oncology, Fundeni Clinical Institute, 258 Fundeni Str., 022238 Bucharest, Romania; vlad.m.croitoru@gmail.com (V.M.C.); irina.cazacu89@gmail.com (I.M.C.)
3 Department of Medical Oncology, University of Medicine and Pharmacy Grigore T Popa Iasi, 700115 Iasi, Romania; simonavolovat@gmail.com
4 Department of Pharmacoeconomics, University of Medicine and Pharmacy of Craiova, 2 Petru Rares Str., 200349 Craiova, Romania; adina.turcu@gmail.com
5 Department of Gastroenterology, University of Medicine and Pharmacy Craiova, 2 Petru Rares Str., 200349 Craiova, Romania; digheonea@gmail.com
6 Department of Medical Oncology, The Oncology Institute "Prof. Dr. Ion Chiricuţă", 400015 Cluj-Napoca, Romania
7 11th Department of Medical Oncology, University of Medicine and Pharmacy "Iuliu Hatieganu", 400012 Cluj-Napoca, Romania
8 Department of Oncology, University of Medicine and Pharmacy Craiova, 2 Petru Rares Str., 200349 Craiova, Romania; cristilungulescu@yahoo.com
* Correspondence: dr.geni@yahoo.co.uk
† These authors contributed equally to this work.

check for updates

Citation: Lungulescu, C.; Croitoru, V.M.; Volovat, S.R.; Cazacu, I.M.; Turcu-Stiolica, A.; Gheonea, D.I.; Sur, D.; Lungulescu, C.V. An Insight into Deficient Mismatch Repair Colorectal Cancer Screening in a Romanian Population—A Bi-Institutional Pilot Study. *Medicina* **2021**, 57, 847. https://doi.org/10.3390/medicina57080847

Academic Editors: Antonio M. Scanu, Konstantinos Dimas and Maria Rosaria De Miglio

Received: 30 June 2021
Accepted: 18 August 2021
Published: 20 August 2021

Publisher's Note: MDPI stays neutral with regard to jurisdictional claims in published maps and institutional affiliations.

Abstract: *Background and Objectives*: Colorectal cancer (CRC) can be classified as mismatch-repair-deficient (dMMR) with high levels of microsatellite instability (MSI-H), or mismatch-repair-proficient (pMMR) and microsatellite stable (MSS). Approximately 15% of patients have microsatellite instability (MSI). MSI-H tumors are associated with a high mutation burden. Monoclonal antibodies that block immune checkpoints can induce long-term durable responses in some patients. Pembrolizumab is the first checkpoint inhibitor approved in the EU to treat dMMR–MSI-H metastatic CRC. *Materials and Methods*: Our study assesses the regional variability of MSI-H colorectal cancer tumors in Romania. Formalin-fixed, paraffin-embedded (FFPE) tissue blocks containing tumor samples from 90 patients diagnosed with colorectal cancer were collected from two tertiary referral Oncology Centers from Romania. Tissues were examined for the expression loss of MMR proteins (MLH1, PMS2, MSH2, MSH6) using immunohistochemistry or MSI status using polymerase chain reaction (PCR), respectively. *Results*: MSI-H was detected in 19 (21.1%) patients. MSI-H was located more in ascending colon (36.8% vs. 9.9%, p-value = 0.0039) and less in sigmoid (5.3% vs. 33.8%, p-value = 0.0136) than MSS patients. Most patients were stage II for MSI-H (42.1%) as well as for MSS (56.3%), with significant more G1 (40.9% vs. 15.8%, p-value = 0.0427) for MSS patients. Gender, N stage, and M stage were identified as significant prognostic factors in multivariate analysis. MSI status was not a statistically significant predictor neither in univariate analysis nor multivariate analysis. *Conclusion*: Considering the efficacy of PD-1 inhibitor in metastatic CRC with MSI-H or dMMR, and its recent approval in EU, it is increasingly important to understand the prevalence across tumor stage, histology, and demographics, since our study displayed higher regional MSI-H prevalence (21%) compared to the literature.

Keywords: MSI; dMMR; colorectal; cancer; immunotherapy; Romania; incidence; PCR; IHC

1. Introduction

As the third most common malignancy and the third leading cause of cancer-related mortality in both genders, colorectal cancer (CRC) poses a severe public health problem globally [1]. However, CRC can be one of the most curable diseases if it is discovered in early settings [2]. Colorectal cancer is a heterogeneous condition generated by the interaction of genetic and environmental components. Molecular variations that occur in CRC can be classified into three major categories: CIN (chromosomal instability), MSI (microsatellite instability), and CIMP (CpG Island Methylator Phenotype)—that causes gene function to be silenced by aberrant hypermethylation [3]. The purpose of this research is to examine MSI instability. Short (1–6 base pair) DNA repeating segments scattered across the entire genome are known as microsatellites or short tandem repeats (STR). Approximately 3 percent of the human genome is comprised of microsatellites, which are vulnerable to mutations due to their repeated structure [4]. An alternate-sized repeating DNA sequence that is not present in germline DNA is the hallmark of microsatellite instability in cancerous cells' DNA. Microsatellite instability (MSI) represents a molecular phenotype caused by a defective DNA mismatch repair system (MMR). During DNA replication and recombination, mistakes such as base-base mismatches and insertions and deletions are corrected by the DNA mismatch repair mechanism. MMR proteins are fundamentally nuclear enzymes thatpromotethe repair of base-base mismatches that arise during cell proliferation by creating complexes (heterodimers) that adhere to aberrant DNA regions and initiate their removal [5]. MMR protein deficiency results in an accumulation of DNA replication defects, particularly in regions of the genome containing short repeating nucleotide sequences, which results in microsatellite instability.

Approximately 15% of patients have microsatellite instability, according to twenty-two relevant publications with sample sizes ranging from 30 to 1000 and data on 7642 patients [5–7]. Three percent of the microsatellite instability-high (MSI-H) tumors have germline mutations in one of the MMR genes, defined as Lynch syndrome [8]. The remaining MSI-H tumors have acquired somatic mutations caused by abnormal methylation of the promoter of a gene that encodes a DNA MMR protein (MLH1).

Lynch syndrome (LS), alternatively referred to as hereditary non-polyposis colorectal cancer (HNPCC), is an inherited autosomal dominant condition that increases the risk of developing certain malignancies, particularly colorectal cancer. This is a consequence of a germline mutation in 1 of several genes involved in DNA mismatch repair (MMR), namely, MLH1, PMS2, MSH2, and MSH6 [9]. Since 90% of colorectal tumors due to LS have microsatellite instability, LS patients and their family members should undergo active surveillance;MSI testing could serve as a screening method.

Our study assesses the regional variability of MSI-H colorectal cancer tumors in Romania, as European Medicines Agency (EMA) recently approved immunotherapy as a treatment for metastatic colorectal cancer patients with high microsatellite instability (MSI-H) or mismatch repair deficiency (dMMR). Studying geographical variations and clinical characteristics of CRC patients is essential since innovative therapies, diagnosis techniques, and new methods of delivering treatments are constantly being developed [10–12].

2. Materials and Methods

Formalin-fixed, paraffin-embedded (FFPE) tissue blocks in which the tumor was visible macroscopically from 90 patients diagnosed with colorectal cancer were collected from two tertiary referral Oncology Centers from Romania. All patients included in this study are ethnic Romanians and of Caucasian descent. All patients received chemotherapy regimens combining fluoropyrimidines and oxaliplatin in an adjuvant setting. Following metastatic disease, targeted therapies such as Cetuximab/Panitumumab or Bevacizumab were added based on KRAS status.

Thirty-three tissue samples were examined for the expression loss of MMR proteins (MLH1, PMS2, MSH2, MSH6) using immunohistochemistry (IHC). Positive staining was confirmed on adjacent normal tissue. MMR protein staining was deemed negative when all cancer cell nuclei failed to react with the antibody (dMMR). Samples missing one or more proteins were considered positive. If all 4 proteins were present, the likelihood of HNPCC/Lynch syndrome is reduced.

Genomic DNA was extracted from the remaining 57 samples after macro-dissection from the cancer tissue, as follows: a certified gastrointestinal pathologist carefully evaluated and dissected the areas of the slides cut from FFPE tissue blocks representing the tumor and "normal" tissue—usually an uninvolved proximal or distal margin of resection. Analysis was carried out using five polymorphic markers (short tandem repeats—STR), referred to as the Bethesda panel, consists of two mononucleotide loci (Big Adenine Tract [*BAT*]-*25* and *BAT-26*) and three dinucleotide loci (*D2S123*, *D5S346*, and *D17S250*) [13]. Using this panel, tumors with instability at two or more of these loci were interpreted as MSI-high. In contrast, the lack of instability at either of the five loci was considered MSS.

Statistical Analysis

Descriptive statistics were generated for all patients. The patients were divided into high microsatellite instability (MSI-H) and microsatellite stability (MSS). We compared the clinical characteristics of patients with MSI-H or MSS. Statistical comparations by microsatellite stability were assessed using Kruskal–Wallis (for continuous variables), Chi-Square (for categorical variables), or Log-rank (Mantel–Cox) test (for progression-free survival, PFS). The survival graph for PFS was generated using the Kaplan–Meier method. Time to event endpoints were analyzed using COX regression. The factors affecting survival (PFS) were identified using univariate and multivariate analysis. Statistical analysis was performed using GraphPad Prism 9.1.2 software (GraphPad Software, San Diego, CA, USA). The power analysis for our study was performed using G*Power 3.1.9.7. A two-sided p-value smaller than 0.05 was considered to be statistically significant.

3. Results

A total of 90 patients were enrolled. MSI-H was detected in 19 (21.1%) patients. Demographic and clinical characteristics of the patients in the MSI-H (n = 19) and MSS (n = 71) are summarized in Table 1. No significant statistical differences in age (p = 0.878) or gender (p = 0.514) were noted between the two groups. Moreover, no significant differences were found for TNM stages.

Table 1. Characteristics of the cohort used in this study.

Variable	Total (n = 90)	Patients MSI-H (n = 19)	MSS (n = 71)	p-Value
Age				
Mean (SD)	61.8 (10.0)	61.8 (10.7)	61.8 (9.9)	0.878 [1]
Median (IQR)	62 (54–67)	62 (52–67)	63 (58–67)	
Range	36–84	45–84	36–83	
Gender, female, n (%)	40 (44.4%)	8 (42.1%)	32 (45.1%)	0.99 [2]
Tumor location, n (%)				
Ascending colon	14 (15.6%)	7 (36.8%)	7 (9.9%)	
Cecum	8 (8.9%)	1 (5.3%)	7 (9.9%)	
Descending colon	12 (13.3%)	3 (15.8%)	9 (12.7%)	
Hepatic angle	1 (1.1%)	0	1 (2.2%)	-
Recto-sigmoid	8 (8.9%)	1 (5.3%)	7 (9.9%)	
Rectum	13 (14.4%)	2 (10.5%)	11 (15.5%)	
Sigmoid	25 (27.8%)	1 (5.3%)	24 (33.8%)	
Superior rectum	3 (3.3%)	1 (5.3%)	2 (2.8%)	
Transverse colon	6 (6.7%)	3 (15.8%)	3 (4.2%)	
Tumor location, n (%)				
Proximal	33 (36.7%)	13 (68.4%)	20 (28.2%)	0.0025 [2,*]
Distal	57 (63.3%)	6 (31.6%)	51 (71.8%)	

Table 1. *Cont.*

Variable	Total (n = 90)	Patients MSI-H (n = 19)	MSS (n = 71)	p-Value
Disease stage				
I	2 (3.6%)	1 (5.3%)	1 (1.4%)	0.447 [2] (I-II vs III-IV)
II	48 (63.6%)	8 (42.1%)	40 (56.3%)	
III	21 (16.4%)	7 (36.8%)	14 (19.7%)	
IV	19 (16.4%)	3 (15.8%)	16 (22.5%)	
Histologic Grade, n (%)				
G1	32 (35.6%)	3 (15.8%)	29 (40.9%)	0.052 [2]
G2	49 (54.4%)	12 (63.2%)	37 (52.1%)	
G3	9 (10.0%)	4 (21.1%)	5 (7%)	
T-Stage, n (%)				
T1	1 (1.1%)	0	1 (1.4%)	0.99 [2] (T1–2 vs. T3–4)
T2	6 (6.7%)	1 (5.3%)	5 (7.0%)	
T3	63 (70.0%)	14 (73.7%)	49 (69.0%)	
T4	13 (14.4%)	4 (21.1%)	9 (12.7%)	
Tx	7 (7.8%)	0	7 (9.9%)	
N-Stage, n (%)				
N0	51 (56.7%)	10 (52.6%)	41 (57.7%)	0.796 [2] (N0 vs N1–2)
N1	18 (20.0%)	4 (21.1%)	14 (19.7%)	
N2	12 (13.3%)	5 (26.3%)	7 (9.9%)	
Nx	9 (10.0%)	0	9 (12.7%)	
M-Stage, n (%)				
M0	70 (77.8%)	17 (89.5%)	53 (74.6%)	0.223 [2] (M0 vs M1)
M1	3 (3.3%)	0	3 (4.2%)	
M1 with hepatic metastases	11 (12.2%)	1 (5.3%)	10 (14.1%)	
M1 with hepatic, pulmonary metastases	3 (3.3%)	1 (5.3%)	2 (2.8%)	
M1 with pulmonary metastases	1 (1.1%)	0	1 (1.4%)	
Mx	2 (2.2%)	0	2 (2.8%)	
Metastatic CRC, yes, n (%)	15 (16.7%)	2 (10.5%)	13 (18.3%)	0.729 [2]

All percentages are based on the total number of patients in each group. CRC, colorectal cancer. [1] Kruskal–Wallis p-value; [2] Fisher's exact test. p-value; *, significant difference.

MSI-H was located more in ascending colon (36.8% vs. 9.9%, p-value = 0.0039) and less in sigmoid (5.3% vs. 33.8%, p-value = 0.0136) than MSS patients. Most patients were stage II for MSI-H (42.1%) as well as for MSS (56.3%), with significant more G1 (40.9% vs. 15.8%, p-value = 0.0427) for MSS patients, as Figure 1 shows.

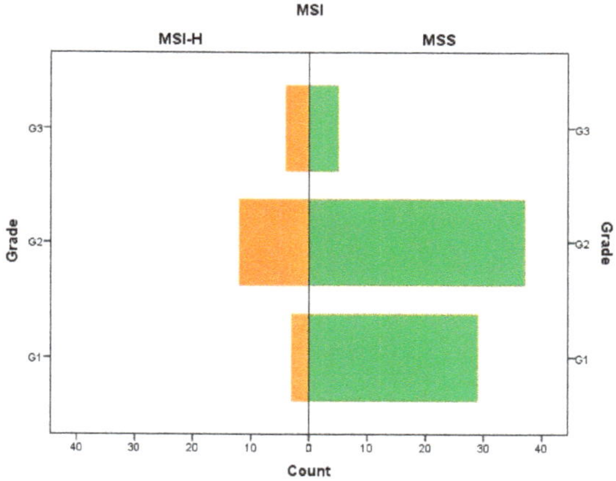

Figure 1. Patients' distribution for histologic grade by MSI.

No difference in PFS was noted between the two groups, as Table 2 and Figure 2 show.

Table 2. PFS comparison between MSI-H and MSS patients.

PFS	Patients			*p*-Value
	Total (n = 55)	MSI-H (n = 9)	MSS (n = 46)	
Mean (SD)	42.2 (27)	38.1 (26.04)	42.9 (27.4)	
Median (IQR)	47.5 (15.3–60.1)	47.3 (13.3–63.9)	47.6 (15.9–60.2)	0.865 [1]
Range	1.9–128.8	6.48–71.5	1.9–128.8	

[1] Log-rank (Mantel–Cox) test *p*-value; PFS, progression-free survival; SD, standard deviation; IQR, interquartile range.

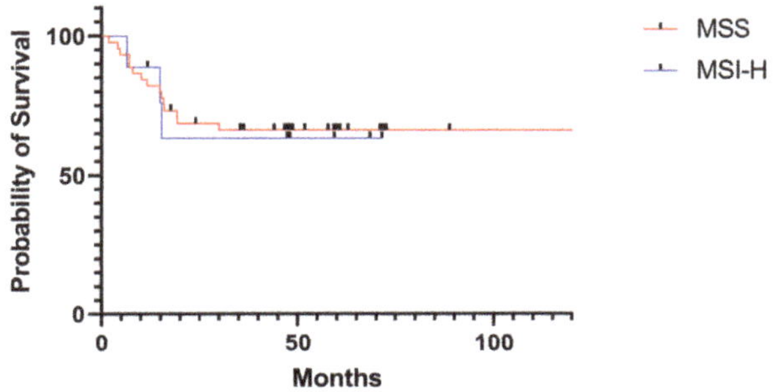

Figure 2. Kaplan– survival curves for patients with colorectal carcinoma by MSI status. MSI, microsatellite instability; MSI-H, high microsatellite instability; MSS, microsatellite stability.

Cox hazard regression was performed to identify the factors affecting the survival rate. In univariate analysis, males, advanced N stage, and M stage were statistically significant predictors of poor outcomes. MSI status didnot point out to be a statistically significant predictor neither in univariate analysis nor multivariate analysis. Gender, N stage, and M stage were identified as significant prognostic factors in multivariate analysis, as in Table 3.

Table 3. Cox proportional hazard regression for clinical characteristics.

Factors		Univariate Analysis		Multivariate Analysis	
		Hazard Ratio	*p*-Value	Hazard Ratio	*p*-Value
Age		1.05 (0.98–1.12)	0.21	-	-
Gender	Male vs. female	30.63 (2.61–359.04)	0.006	5.33 (1.24–22.98)	0.025
MSI	MSS vs. MSI-H	8.24 (0.92–73.54)	0.059	3.14 (0.7–14.16)	0.136
Location	Proximal vs. distal	3.76 (0.55–25.53)	0.175	-	-
Stage	III-IV vs. I-II	0.09 (0.01–1.59)	0.1	-	-
T stage	T3–4 vs. T1–2	2.47 (0.32–19.2)	0.388	-	-
N stage	N0 vs. N1–2	0.01 (0.0–0.53)	0.023	0.01 (0.001–0.104)	<0.001
M stage	M0 vs. M1	0.25 (0.07–0.91)	0.036	0.19 (0.05–0.7)	0.012

We calculated the power for the outcomes from our study, and the obtained value of 90.8% demonstrates the sample size is representative ofour study's results.

4. Discussion

Alteration in MMR proteins is frequently associated with the absence of an identifiable gene product, enabling IHC testing to indirectly determine the expression loss of the respective genes. IHC displays certain advantages over MSI analysis, such as relatively inexpensive and routinely used techniques. Moreover, it offers gene-specific information—the absence of a certain mismatch gene product (MLH1, MSH2, MSH6, or PMS2) can guide germline testing and aid in identifying patients with LS. However, IHC is susceptible to the quality of tissue preparation, variability of the antibodies, and interpretation—not a standardized method. Studies suggest that MSI testing and IHC are complimentary, as the loss of MMR protein expression is highly concordant with DNA-based MSI testing, providing a good sensitivity (>90%) and excellent specificity (100%) [14]. Several investigations revealed nearly perfect concordance between PCR and IHC tests. As a result, either approach is appropriate as a first-line screening method for determining dMMR/MSI-H status [15]. Additional criteria, such as more comprehensive family histories and genetic tests, including BRAF V600E mutation and hypermethylation of the MLH1 promoter, are necessary to differentiate between sporadic and hereditary colorectal cancer [16].

Our data showed noticeably higher regional MSI-H prevalence (21%) compared to other populations (10–13%) [4,5]. MSI-H was located more in ascending colon (36.8% vs. 9.9%, p-value = 0.0039) and less in the sigmoid (5.3% vs. 33.8%, p-value = 0.0136) than MSS patients, as reported in previous studies that have described that MSI-H is more frequently observed in proximal colon tumors than distal colon cancers.

The results of our study showed that gender, N stage, and M stage were identified as significant prognostic factors in multivariate analysis. These results support the notion that the TNM stage prevails as the gold standard for diagnosing colorectal tumors. However, numerous retrospective and population-based investigations have demonstrated that patients with dMMR tumors had a better stage-adjusted prognosis, implying that the superior outcomeassociated with dMMR CRCs is more apparent in early-stage lesions [7,17].

Patients with MSI-H CRC had a better prognosis, although it is unclear if MSI status predicts responsiveness to adjuvant chemotherapy [18]. Neither the univariate nor the multivariate analysis in the present study suggested that the MSI status significantly influenced prognosis.

The differentresponse of MSI-H tumors to chemotherapeutic drugs has been extensively studied in experiments. In addition to alkylating agents, DNA dMMR cells are resistant to platinum-containing treatments (cisplatin and carboplatin), antimetabolites (fluorouracil), and topoisomerase inhibitors (doxorubicin) [19]. According to findings, patients with stage II or stage III CRC with MSS tumors benefited from fluorouracil-based adjuvant treatment [20]. Cellular dynamics linked with MMR downregulation may explain these outcomes (increased apoptosis and decreased proliferation). However, multiple studies have been published that suggest MSI-H as a predictor of enhanced response to irinotecan or irinotecan-based chemotherapy in CRC patients.

Additionally, to quantify the response to chemotherapy, MSI has been recently established as a major predictive marker for immune checkpoint blockade response. Antitumor immune responses within MSI tumors are stronger than their MSS counterparts due to the high tumor mutational burden and neoantigen load that promote the infiltration of immune effector cells [21].

Nivolumab plus low-dose ipilimumab was authorized by the U.S. Food and Drug Administration on 11 July 2018, to treat MSI-H or dMMR metastatic colorectal cancer that has progressed after treatment with fluoropyrimidine, oxaliplatin, and irinotecan. The approval was based on findings from the Check-Mate-142 phase II investigation [22]. Subsequently, on 29 June 2020, the FDA approved pembrolizumab for the first-line treatment of patients with unresectable or metastatic colorectal cancer with high microsatellite instability or mismatch repair deficiency. The indication was approved based on the results of KEYNOTE 177 (NCT02563002), a trial in which 307 patients with previously untreated unresectable or metastatic MSI-H or dMMR colorectal cancer were included [23]. Similarly, on 10 December 2020, the European Medicines Agency's (EMA's) Committee for Medicinal Products for Human Use (CHMP) adopted the new indication for the medicinal product pembrolizumab. This marks the first approval of the CHMP for a target population defined by DNA repair deficiency biomarkers [24]. Pembrolizumab is recommended at a dose of 200 mg every three weeks or 400 mg every six weeks for MSI-H/dMMR colorectal cancer. Recently, on 20 May 2021, CHMP recommended nivolumab in combination with ipilimumab for the treatment of MSI-H/dMMR metastatic colorectal cancer patients who had previously received fluoropyrimidine-based combination therapy [25].

However, our findings show that MSI-H tumors had a lower rate of metastatic disease (10.5%) than MSS CRC (18.3%), highlighting the importance of future prospective large trials to demonstrate immunotherapy's relevance in the adjuvant or neoadjuvant setting for CRC. FOLFOX6 adjuvant treatment with or without atezolizumab is currently being evaluated in phase III ATOMIC study (NCT02912559) to assess if the combined therapies offer a higher survival benefit than conventional chemotherapy alone for stage III dMMR CRC [26].

5. Conclusions

MSI is a surrogate marker of DNA mismatch repair deficiency and a surrogate for neoantigen load that enhances antitumor immune response. In light of evidence supporting the efficacy of PD-1 inhibitor in metastatic CRC with MSI-H or dMMR, and its recent approval in the EU, it is increasingly important to understand the prevalence across tumor stage, histology, and demographics, since our study displayed higher regional MSI-H prevalence (21%) compared to the literature.

Author Contributions: Conceptualization, C.L. and C.V.L.; methodology, S.R.V.; validation, D.S. and I.M.C.; performed the analysis, and designed the figures, A.T.-S.; resources, C.L. and V.M.C.; writing—original draft preparation, C.L. and V.M.C.; writing—review and editing, D.S. and C.V.L.; supervision, D.I.G. All authors have read and agreed to the published version of the manuscript.

Funding: This research received no external funding.

Institutional Review Board Statement: Ethical review and approval were waived for this study because the research presents no risk of harm to participants and involves no procedures for which written consent is normally required outside of the research context. Furthermore, the identifiable private information or identifiable biospecimens are not expressed.

Informed Consent Statement: Not applicable.

Data Availability Statement: The datasets used and/or analyzed during the current study are available from the corresponding author on reasonable request.

Conflicts of Interest: The authors declare no conflict of interest.

References

1. Bray, F.; Ferlay, J.; Soerjomataram, I.; Siegel, R.L.; Torre, L.A.; Jemal, A. Global cancer statistics 2018: GLOBOCAN estimates of incidence and mortality worldwide for 36 cancers in 185 countries. *CA Cancer J. Clin.* **2018**, *68*, 394–424. [CrossRef]
2. Popa, P.; Streba, C.T.; Caliţă, M.; Iovănescu, V.F.; Florescu, D.N.; Ungureanu, B.S.; Stănculescu, A.D.; Ciurea, R.N.; Oancea, C.N.; Georgescu, D.; et al. Value of endoscopy with narrow-band imaging and probe-based confocal laser endomicroscopy in the diagnosis of preneoplastic lesions of gastrointestinal tract. *Rom. J. Morphol. Embryol.* **2020**, *61*, 759–767. [CrossRef] [PubMed]
3. Worthley, D.L.; Leggett, B.A. Colorectal cancer: Molecular features and clinical opportunities. *Clin. Biochem. Rev.* **2010**, *31*, 31–38.
4. Ellegren, H. Microsatellites: Simple sequences with complex evolution. *Nat. Rev. Genet.* **2004**, *5*, 435–445. [CrossRef]
5. Boland, C.R.; Goel, A. Microsatellite instability in colorectal cancer. *Gastroenterology* **2010**, *138*, 2073–2087.e3. [CrossRef] [PubMed]
6. Vilar, E.; Gruber, S.B. Microsatellite instability in colorectal cancer-the stable evidence. *Nat. Rev. Clin. Oncol.* **2010**, *7*, 153–162. [CrossRef]
7. Popat, S.; Hubner, R.; Houlston, R.S. Systematic review of microsatellite instability and colorectal cancer prognosis. *J. Clin. Oncol.* **2005**, *23*, 609–618. [CrossRef]
8. Hampel, H.; Frankel, W.L.; Martin, E.; Arnold, M.; Khanduja, K.; Kuebler, P.; Nakagawa, H.; Sotamaa, K.; Prior, T.W.; Westman, J.; et al. Screening for the Lynch syndrome (hereditary nonpolyposis colorectal cancer). *N. Engl. J. Med.* **2005**, *352*, 1851–1860. [CrossRef]
9. Lynch, H.T.; Snyder, C.L.; Shaw, T.G.; Heinen, C.D.; Hitchins, M.P. Milestones of Lynch syndrome: 1895–2015. *Nat. Rev. Cancer* **2015**, *15*, 181–194. [CrossRef]
10. Cisterna, B.A.; Kamaly, N.; Choi, W.I.; Tavakkoli, A.; Farokhzad, O.C.; Vilos, C. Targeted nanoparticles for colorectal cancer. *Nanomedicine* **2016**, *11*, 2443–2456. [CrossRef] [PubMed]
11. Turcu-Stiolica, A.; Popescu, M.; Bubulica, M.; Oancea, C.; Nicolicescu, C.; Manda, C.; Neamtu, J.; Croitoru, O. Optimization of Gold Nanoparticles Synthesis using Design of Experiments Technique. *Rev. Chimie.* **2017**, *68*, 1518–1523. [CrossRef]
12. Şerbănescu, M.S.; Oancea, C.N.; Streba, C.T.; Pleşea, I.E.; Pirici, D.; Streba, L.; Pleşea, R.M. Agreement of two pre-trained deep-learning neural networks built with transfer learning with six pathologists on 6000 patches of prostate cancer from Gleason2019 Challenge. *Rom. J. Morphol. Embryol.* **2020**, *61*, 513–519. [CrossRef]

13. Berg, K.D.; Glaser, C.L.; Thompson, R.E.; Hamilton, S.R.; Griffin, C.A.; Eshleman, J.R. Detection of microsatellite instability by fluorescence multiplex polymerase chain reaction. *J. Mol. Diagn.* **2000**, *2*, 20–28. [CrossRef]
14. Lindor, N.M.; Burgart, L.J.; Leontovich, O.; Goldberg, R.M.; Cunningham, J.M.; Sargent, D.J.; Walsh-Vockley, C.; Petersen, G.M.; Walsh, M.D.; Leggett, B.A.; et al. Immunohistochemistry versus microsatellite instability testing in phenotyping colorectal tumors. *J. Clin. Oncol.* **2002**, *20*, 1043–1048. [CrossRef] [PubMed]
15. Loughrey, M.B.; McGrath, J.; Coleman, H.G.; Bankhead, P.; Maxwell, P.; McGready, C.; Bingham, V.; Humphries, M.P.; Craig, S.G.; McQuaid, S.; et al. Identifying mismatch repair-deficient colon cancer: Near-perfect concordance between immunohistochemistry and microsatellite instability testing in a large, population-based series. *Histopathology* **2021**, *78*, 401–413. [CrossRef] [PubMed]
16. Kunnackal John, G.; Das Villgran, V.; Caufield-Noll, C.; Giardiello, F. Worldwide variation in lynch syndrome screening: Case for universal screening in low colorectal cancer prevalence areas. *Fam. Cancer* **2021**, *20*, 145–156. [CrossRef] [PubMed]
17. Roth, A.D.; Delorenzi, M.; Tejpar, S.; Yan, P.; Klingbiel, D.; Fiocca, R.; d'Ario, G.; Cisar, L.; Labianca, R.; Cunningham, D.; et al. Integrated analysis of molecular and clinical prognostic factors in stage II/III colon cancer. *J. Natl. Cancer Inst.* **2012**, *104*, 1635–1646. [CrossRef]
18. Kawakami, H.; Zaanan, A.; Sinicrope, F.A. Microsatellite instability testing and its role in the management of colorectal cancer. *Curr. Treat Options Oncol.* **2015**, *16*, 30. [CrossRef]
19. Damia, G.; D'Incalci, M. Genetic instability influences drug response in cancer cells. *Curr. Drug Targets* **2010**, *11*, 1317–1324. [CrossRef]
20. Ribic, C.M.; Sargent, D.J.; Moore, M.J.; Thibodeau, S.N.; French, A.J.; Goldberg, R.M.; Hamilton, S.R.; Laurent-Puig, P.; Gryfe, R.; Shepherd, L.E.; et al. Tumor microsatellite-instability status as a predictor of benefit from fluorouracil-based adjuvant chemotherapy for colon cancer. *N. Engl. J. Med.* **2003**, *349*, 247–257. [CrossRef]
21. Drescher, K.M.; Sharma, P.; Watson, P.; Gatalica, Z.; Thibodeau, S.N.; Lynch, H.T. Lymphocyte recruitment into the tumor site is altered in patients with MSI-H colon cancer. *Fam. Cancer* **2009**, *8*, 231–239. [CrossRef] [PubMed]
22. Overman, M.J.; McDermott, R.; Leach, J.L.; Lonardi, S.; Lenz, H.J.; Morse, M.A.; Desai, J.; Hill, A.; Axelson, M.; Moss, R.A.; et al. Nivolumab in patients with metastatic DNA mismatch repair-deficient or microsatellite instability-high colorectal cancer (CheckMate 142): An open-label, multicentre, phase 2 study. *Lancet Oncol.* **2017**, *18*, 1182–1191. [CrossRef]
23. André, T.; Shiu, K.K.; Kim, T.W.; Jensen, B.V.; Jensen, L.H.; Punt, C.; Smith, D.; Garcia-Carbonero, R.; Benavides, M.; Gibbs, P.; et al. KEYNOTE-177 Investigators. Pembrolizumab in Microsatellite-Instability-High Advanced Colorectal Cancer. *N. Engl. J. Med.* **2020**, *383*, 2207–2218. [CrossRef] [PubMed]
24. Trullas, A.; Delgado, J.; Genazzani, A.; Mueller-Berghaus, J.; Migali, C.; Müller-Egert, S.; Zander, H.; Enzmann, H.; Pignatti, F. The EMA assessment of pembrolizumab as monotherapy for the first-line treatment of adult patients with metastatic microsatellite instability-high or mismatch repair deficient colorectal cancer. *ESMO Open* **2021**, *6*, 100145. [CrossRef]
25. Bristol Myers Squibb. Bristol Myers Squibb Receives Positive CHMP Opinion recommending Opdivo (nivolumab) Plus Yervoy (Ipilimumab) for Treatment of Mismatch Repair Deficient or Microsatellite Instability–High Metastatic Colorectal Cancer after Prior Chemotherapy. *News Release*. 21 May 2021. Available online: https://bit.ly/3wsujTG (accessed on 24 May 2021).
26. Sinicrope, F.A.; Ou, F.S.; Zemla, T.; Nixon, A.B.; Mody, K.; Levasseur, A.; Dueck, A.C.; Dhanarajan, A.R.; Hanyoung Lieu, C.; Cohen, D.J.; et al. Randomized trial of standard chemotherapy alone or combined with atezolizumab as adjuvant therapy for patients with stage III colon cancer and deficient mismatch repair (ATOMIC, Alliance A021502). *JCO* **2019**, *37* (Suppl. 15), e15169. [CrossRef]

medicina

MDPI

Article

Receptor-Interacting Serine/Threonine-Protein Kinase-2 as a Potential Prognostic Factor in Colorectal Cancer

Rola F. Jaafar [1], Zeid Ibrahim [1], Karim Ataya [2], Joelle Hassanieh [1], Natasha Ard [3] and Walid Faraj [1,2,*]

[1] Department of Surgery, American University of Beirut Medical Center, Beirut 1107 2020, Lebanon; rj29@aub.edu.lb (R.F.J.); iz13@aub.edu.lb (Z.I.); jh91@aub.edu.lb (J.H.)
[2] Division of Liver Transplantation, Hepatobiliary and Pancreatic Surgery, Department of General Surgery, American University of Beirut Medical Centre, Beirut 1107 2020, Lebanon; ka71@aub.edu.lb
[3] Department of General Medicine, Faculty of Medicine, American University of Beirut Medical Center, Beirut 1107 2020, Lebanon; noa09@aub.edu.lb
* Correspondence: wf07@aub.edu.lb; Tel.: +961-350-000 (ext. 5714)

Abstract: *Background and objectives*: Receptor-interacting serine/threonine-protein kinase-2 (RIPK2) is an important mediator in different pathways in the immune and inflammatory response system. RIPK2 was also shown to play different roles in different cancer types; however, in colorectal cancer (CRC), its role is not well established. This study aims at identifying the role of RIPK2 in CRC progression and survival. *Materials and methods*: Data of patients and mRNA protein expression level of genes associated with CRC (RIPK2, tumor necrosis factor (TNF), TRAF1, TRAF7, KLF6, interlukin-6 (Il6), interlukin-8 (Il8), vascular-endothelial growth factor A (VEGFA), MKI67, TP53, nuclear factor-kappa B (NFKB), NFKB2, BCL2, XIAP, and RELA) were downloaded from the PrognoScan online public database. Patients were divided between low and high RIPK2 expression and different CRC characteristics were studied between the two groups. Survival curves were evaluated using a Kaplan–Meier estimator. The Pearson correlation was used to study the correlation between RIPK2 and the other factors. Statistical analysis was carried out using SPSS version 25.0. The Human Protein Atlas was also used for the relationship between RIPK2 expression in CRC tissues and survival. Differences were considered statistically significant at $p < 0.05$. *Results*: A total of 520 patients were downloaded from the PrognoScan database, and RIPK2 was found to correlate with MKI67, TRAF1, KLF6, TNF, Il6, Il8, VEGFA, NFKB2, BCL2, and RELA. High expression of RIPK2 was associated with high expression of VEGFA ($p < 0.01$) and increased mortality ($p < 0.01$). *Conclusions*: In this study, RIPK2 is shown to be a potential prognostic factor in CRC; however, more studies are needed to assess and verify its potential role as a prognostic marker and in targeted therapy.

Keywords: colorectal cancer; inflammatory pathways; NFKB; RIPK2

Citation: Jaafar, R.F.; Ibrahim, Z.; Ataya, K.; Hassanieh, J.; Ard, N.; Faraj, W. Receptor-Interacting Serine/Threonine-Protein Kinase-2 as a Potential Prognostic Factor in Colorectal Cancer. *Medicina* **2021**, *57*, 709. https://doi.org/10.3390/medicina57070709

Academic Editors: Antonio M. Scanu, Maria Rosaria De Miglio and Andrea Mingoli

Received: 15 April 2021
Accepted: 3 June 2021
Published: 14 July 2021

1. Introduction

The role of inflammation in promoting cancer cell progression is currently a well-known phenomenon [1,2]. Cancer tissues show signs of inflammation, such as the presence of immune cells in the tissue, presence of specific chemokines, and angiogenesis [3]. Chronic inflammation causes tissue damage, which induces cell proliferation and tissue repair and, as a consequence, tumor development [4–6].

Receptor-interacting serine/threonine-protein kinase-2 (RIPK2) is an important mediator required in different pathways in the immune and inflammatory response system, and was found to be involved in different solid tumors [7,8]. It is highly expressed in head and neck squamous cell carcinoma (HNSCC), and it was reported to promote cell proliferation and prevent apoptosis in glioma [9,10]. In breast cancer, mainly in triple-negative breast cancer (TNBC), it was shown to impact patient overall survival, increase recurrence, protect cells from apoptosis induced by chemotherapy, and enhance cell proliferation by activating nuclear factor-kappa B (NFKB) [11–13]. RIP2 expression correlated with the tumor size, metastasis, overall staging, progression-free survival, and body mass index (BMI) of patients with breast cancer, and RIPK2 polymorphism was also involved in the development of bladder cancer [14,15]. In addition, RIPK2 overexpression is associated with cell proliferation and progression of gastric cancer, and is frequently amplified in lethal prostate cancers, leading to disease progression and aggressiveness [16,17].

Inflammation is known to be an essential tumorigenic factor in colorectal cancer (CRC), and several markers have been suggested to play a role in CRC mediation [18]. Strong evidence suggests that NFKB-mediated inflammation is a key element in the etiology of CRC [19]. RIPK2 is associated with the NFKB pathway and seems to have a role in colitis-associated CRC, where the level of expression of RIPK2 was significantly higher in the colonic mucosa of patients with ulcerative colitis compared to controls [20]. In mice, a deficiency in RIPK2 can cause dysbiosis, which is a microbial imbalance in the colon, which in turn predisposes mice to communicable colitis and colitis-associated CRC [21]. In a recent study on four patients with CRC, RIPK2 was shown to be upregulated in rectal cancer in comparison to normal adjacent mucosa, as identified by the ChIP-Seq procedure [22].

Moreover, RIPK2 expression was reported to be associated with the expression of proto-oncogenic proteins including proliferation marker KI67 (MKI67), tumor protein P53 (TP53), and vascular endothelial growth factor A (VEGFA) [14,15]. These proteins also play a role in the survival and prognosis of CRC patients, where high MKI67 expression is correlated with decreased overall survival and disease-free survival, TP53 expression was found to be significantly associated with poor survival, and VEGF expression was associated with decreased survival, higher grade, presence of lymph node metastasis, depth of invasion, and overall stage [23–28].

As mentioned, RIPK2 overexpression is associated with several solid tumors acting through activation of the NFKB and inflammatory pathways. Therefore, this study aims at identifying the role of RIPK2 in CRC patients' prognosis and survival and its association with the different pro-survival proteins involved in CRC, presenting RIPK2 as a potential biomarker and a therapeutic target.

2. Materials and Methods

mRNA proteins' expression in CRC were downloaded from PrognoScan online public database [29]. mRNA expression level of genes associated with CRC tumorigenesis including RIPK2 and defined proto-oncogenic proteins MKI67, TP53, and VEGF, in addition to tumor necrosis factor (TNF), TNF receptor-associated factor 1 (TRAF1), TRAF7, kruppel-like factor 6 (KLF6), interleukin-6 (Il6), interleukin-8 (Il8), NFKB, NFKB2, B-cell lymphoma 2 (Bcl2), X-linked inhibitor of apoptosis protein (XIAP), and v-rel reticuloendotheliosis viral oncogene homolog A (RELA) were analyzed. Data available included patients' age, gender, follow-up time, CRC site, histological markers including grade, and TNM staging [30]. Data were downloaded and entered into Statistical Package for the Social Sciences (SPSS) version 25.0 for analysis.

Data was analyzed based on RIPK2 expression as low or high relative to the mean. RIPK2 expression in each dataset was normally distributed. After removal of outliers, RIPK2 mean was determined in each database as a cut-off value and RIPK2 expression above the mean was considered as high, and expression below the mean was considered as low. Additionally, the expression of MKI67, TP53, and VEGF was normally distributed, and low and high expression were defined based on the mean.

Datasets were combined to analyze the association between low and high expression of RIPK2 and different CRC characteristics. Survival curves were evaluated using the Kaplan–Meier estimators. Pearson correlation was used in each dataset separately to study the correlation between RIPK2 and the other proteins. Differences were considered statistically significant at $p < 0.05$.

3. Results

A total of four databases with a total of 520 patients were obtained from the PrognoScan database and are detailed in Table 1.

Table 1. CRC PrognoScan dataset characteristics.

Dataset	Cohort	Contributor	Year	n	Array Type	Age (Mean ± SD)
GSE12945	Berlin	Staub	2009	62	HG-U133A	64.45 ± 11.78
GSE17536	MCC	Smith	2009	177	HG-U133_Plus_2	65.48 ± 13.08
GSE1433	Melbourne	Jorissen	2010	226	HG-U133_Plus_2	66.03 ± 13.01
GSE17537	VMC	Smith	2009	55	HG-U133_Plus_2	62.31 ± 14.35

3.1. RIPK2 Expression Is Associated with Tumor Site and Grade

Comparing patients based on RIPK2 expression, 273 (52.5%) patients were found to have low expression and 247 (47.5%) had high expression (Table 2). The mean age was 65.54 ± 12.40 and 64.96 ± 13.77, respectively. Higher RIPK2 expression was observed in 16 (59.2%) patients with colon cancer and 11 (40.7%) patients with rectal cancer ($p = 0.08$). Higher expression of RIPK2 was also associated with an increased proportion of grade 3 tumors in 36 patients (27.7%) compared to lower expression in 25 patients (17.2%) ($p = 0.09$). There was no statistically significant association between RIPK2 expression and lymph node involvement, metastasis, and overall stage (Table 2).

Table 2. Patients' characteristics according to RIPK2 expression level.

Variable	Low Expression of RIPK2 (n = 273)	High Expression of RIPK2 (n = 247)	p
Age (mean ± SD)	65.54 ± 12.40	64.96 ± 13.77	0.61
Sex			
Male	146 (53.4%)	130 (52.6%)	0.85
Female	127 (46.6%)	117 (47.4%)	
Site			
Colon	13 (37.1%)	16 (59.3%)	0.08
Rectum	22 (62.9%)	11 (40.7%)	
Tumor grade			
1	11 (7.6%)	6 (4.6%)	0.09
2	109 (75.2%)	88 (67.7%)	
3	25 (17.2%)	36 (27.7%)	
T			
2	9 (25.7%)	7 (25.9%)	0.41
3	25 (71.4%)	17 (62.9%)	
4	1 (2.8%)	3 (11.1%)	
N			
0	22 (62.8%)	14 (51.8%)	0.50
1	6 (17.1%)	8 (29.6%)	
2	7 (20.0%)	5 (18.5%)	
M			
0	30 (88.2%)	26 (96.3%)	0.25
1	4 (11.8%)	1 (3.7%)	
Stage			
I	51 (18.7%)	31 (12.5%)	
II	100 (36.6%)	89 (36.0%)	0.19
III	90 (32.9%)	98 (39.7%)	
IV	32 (11.7%)	29 (11.7%)	

3.2. RIPK2 Association with Proto-Oncogenes

An association between RIPK2 and other proto-oncogenic proteins' mRNA expression was obtained (Table 3). The RIPK2 association with high MKI67 mRNA expression was moderate ($p = 0.06$) and significant with high VEGFA mRNA expression ($p < 0.01$), while no significant association was found with TP53 expression.

Table 3. Association of RIPK2 with proliferation genes' expression.

High Expression of Gene	Low Expression of RIPK2 (n = 273)	High Expression of RIPK2 (n = 247)	p
MKI67	139 (47.2%)	141 (57.0%)	0.06
TP53	166 (60.8%)	155 (62.7%)	0.65
VEGFA	116 (42.4%)	139 (56.2%)	<0.01

A correlation analysis was also carried out to understand the relationship between the expression of all the basic proteins mentioned above and RIPK2 expression (Table 4). MKI67, TRAF1, KLF6, Il6, Il8, VEGFA, and RELA were found to positively correlate with RIPK2 and the highest correlation was found between RIPK2 and RELA in the GSE12945 dataset (k = 0.43; $p < 0.01$). TNF and BcL2 mRNA expression negatively correlated with RIPK2 in one dataset each with $p < 0.01$. However, for NFKB2, it was shown to positively correlate with RIPK2 in the GSE12945 dataset (k = 0.42, $p < 0.01$), but a negative correlation was found in another dataset, GSE17537 (k = −0.30, $p \leq 0.05$). TP53, NFKB, XIAP, and TRAF7 mRNA expression were not significantly correlated with RIPK2 mRNA expression in any of the datasets.

Table 4. Correlation between RIPK2 mRNA expression and proteins involved in colorectal cancer.

Dataset	MKI67	TRAF1	KLF6	TNF	Il6	Il8	VEGFA	NFKB2	BCL2	RELA
GSE12945	0.16	0.15	0.13	0.09	0.02	0.25 *	0.18	0.42 **	0.22	0.43 **
GSE17536	0.10	−0.03	0.24 **	0.12	0.22 **	0.41 **	0.19 *	−0.06	−0.27 **	0.12
GSE1433	0.15 *	0.16 *	0.20 **	−0.00	0.25 **	0.04	0.20 **	0.12	0.01	0.15 **
GSE17537	0.40 **	0.00	0.12	−0.35 **	0.12	−0.08	−0.00	−0.30 *	−0.19	−0.01

* $p \leq 0.05$; ** $p < 0.01$.

3.3. Effect of RIPK2 on the Survival of CRC Patients

Patients with higher RIPK2 mRNA expression had a significantly higher mortality rate (94 patients, 38.06%) in comparison to those with relatively lower expression (61 patients, 22.34%) ($p < 0.01$). Survival analysis was also carried out based on RIPK2 mRNA expression, where high expression of RIPK2 was associated with decreased survival in CRC patients (Figure 1).

Sub-analysis of two databases (GSE12945 and GSE1433) reporting tumor grade and stage was carried out based on survival and is detailed in Table 5. Survival outcome was associated with tumor stage and RIPK2 expression. Moreover, RIPK2 expression was associated with tumor stage, where CRC stage 3 and 4 had significantly more RIPK2 expression relative to stages 1 and 2 (Figure 2).

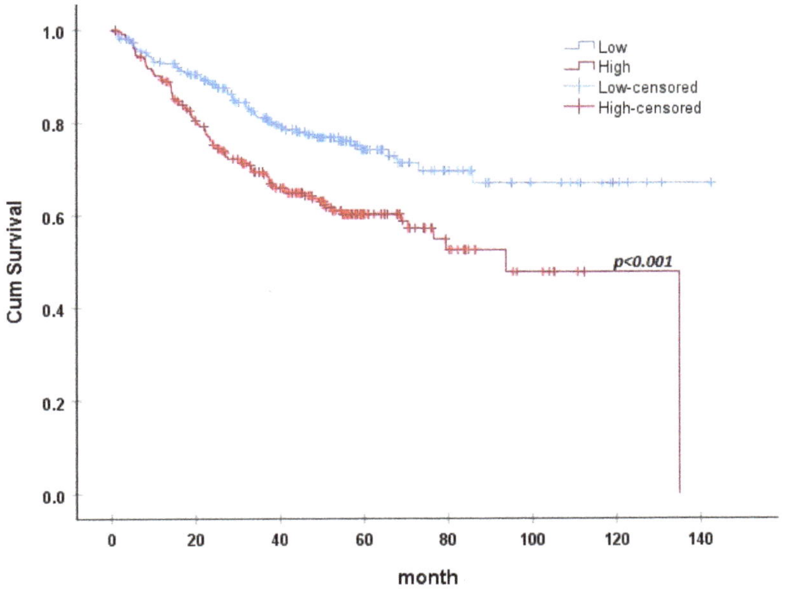

Figure 1. Kaplan–Meier survival plot of patients based on RIPK2 expression.

Table 5. Sub-analysis of two databases reporting tumor stage and grade.

Variable	Event (n = 99)	No Event (n = 176)	p
Age (mean ± SD)	65.98 ± 13.98	64.48 ± 12.04	0.35
Sex			
Male	53 (53.0%)	96 (53.3%)	0.95
Female	47 (47.0%)	84 (46.7%)	
Site			
Colon	6 (50.0%)	23 (46.0%)	0.80
Rectum	6 (50.0%)	27 (54.0%)	
Tumor grade			
1	4 (4.0%)	13 (7.4%)	0.20
2	68 (68.7%)	129 (73.3%)	
3	27 (19.3%)	34 (27.3%)	
Stage			
I	5 (5.0%)	36 (20.0%)	<0.01
II	17 (17.0%)	73 (40.6%)	
III	33 (33.0%)	60 (33.3%)	
IV	45 (45.0%)	11 (6.1%)	
TP53	61 (61.0%)	110 (61.1%)	0.98
VEGF-a	54 (54.0%)	77 (42.8%)	0.07
MKI67	51 (51.0%)	93 (51.7%)	0.91
RIPK2	57 (57.0%)	75 (41.7%)	0.014

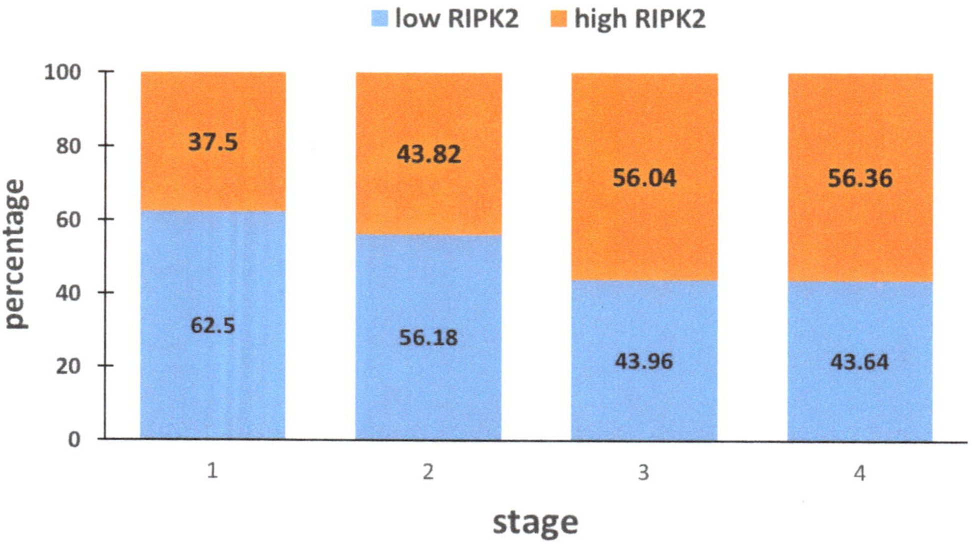

Figure 2. RIPK2 expression relative to tumor stage.

4. Discussion

CRC is the third most common cancer and accounts for 10% of all annually diagnosed cancer, and 9% of cancer-related deaths [31]. Prognostic factors of CRC include the TNM staging, based on which therapeutic decisions are made, and many potential molecular prognostic markers have been described; unfortunately, most have very limited value in routine clinical practice [32].

Surgery, chemotherapy, radiotherapy, and targeted therapy are all options in metastatic CRC. Target therapy includes anti-angiogenetic factors such as bevacizumab, anti-epidermal growth factor receptor-like cetuximab and panitumumab, immune checkpoint inhibitors, anti-BRAF therapy, and HER2-targeted therapy [33]. However, the use of these targeted therapies is limited to factors unique to each patient; for example, the effectiveness of cetuximab is limited to patients with KRAS wild-type tumors, and recent studies also showed that the side of the primary tumor affects the outcome of treatment with cetuximab, with the left-side location being more favorable [34,35]. Hence, discovering other prognostic factors, possible to be targeted by therapy, can improve the outcome of CRC.

We found that the level of mRNA expression of RIPK2 significantly correlated with several proteins involved in tumorigenesis, and after dividing the patients between high and low expression, those who had a higher expression of RIPK2 also had a higher expression of VEGFA ($p < 0.01$). In addition, single database analysis showed a positive correlation between RIPK2 expression and different markers that are established to have a role in CRC, mainly Il-6, Il-8, and VEGF (Table 4), which indicates a potential involvement of RIPK2 in driving tumorigenesis. However, this correlation was not consistent among the different databases, which might be due to the number of patient samples or the data type. Hence, further studies are needed to establish this association. Higher expression of RIPK2 was also associated with worse survival ($p < 0.01$), which suggests RIPK2 as a potential prognostic marker in CRC.

Exploring the Human Protein Atlas, immunohistochemistry analysis was carried out for patients with colon cancer and those with rectal cancer, even though no results were significant, but the patients with rectal cancer were the only ones who showed results similar to our analysis, in which patients with higher expression of RIPK2 showed more death (data not shown). From the data collected from PrognoScan, only 62 patients (11.92%)

had information about the site of cancer, so we were not able to carry out separate analysis by site, but with the results shown in the Human Protein Atlas, we can assume that the majority of our patients had rectal cancer.

Grade and the presence of metastasis are important prognostic factors in CRC, and in our analysis based on the PrognoScan data, RIPK2 expression was associated neither with grade ($p = 0.09$) nor with the presence of metastasis ($p = 0.25$), but it was associated with long-term survival ($p < 0.01$). This is basically due to the nature of the data and the availability of tumor grade in only two datasets, which also limited our ability to perform multivariate analysis. However, when analyzing the datasets reporting tumor stage (GSE12945 and GSE1433), a significant association was observed between high RIPK2 expression and tumor grade, which indicates that RIPK2 is a potential prognostic marker, but more large studies and databases reporting tumor characteristics are needed. However, another limit to our study is that information regarding grade and the presence of metastasis was only present for 275 (52.88%) and 61 (11.73%) patients, respectively. Furthermore, no data was available describing the therapy regimen taken by the patients, and hence the association between RIPK2 and survival based on therapy cannot be reported in these datasets. However, some studies showed that RIPK2 plays a role in metastasis in different cancer forms; for example, in TNBC, RIPK2 knockdown decreases migration and lung metastasis, in inflammatory breast cancer, higher RIPK2 activity was correlated with metastasis, and in hepatic cell carcinoma, knockdown of RIPK2 downregulated multiple genes involved in epithelial–mesenchymal transition [12,14,36].

Moreover, data concerning the relationship between RIPK2 and recurrence of CRC was not available in the studied datasets. It was shown that in TNBC, higher expression of RIPK2 is associated with increased recurrence; hence, further studies should be conducted to determine the role of RIPK2 in metastasis and recurrence in CRC, to see if it differs depending on the site of the tumor, and to assess its status as a potential target for therapy in metastatic CRC [13]. So far there is no targeted therapy for RIPK2 in cancer in general and, hence, in CRC; however, several RIPK2 inhibitors have been under investigation in several inflammatory diseases [37]. However, RIPK2 inhibitors are promising therapeutic drugs for cancer, especially inflammation-associated cancer. Therefore, further studies identifying RIPK2 as a prognostic factor and tumor marker in CRC entails a potential hope for CRC targeted therapy.

Author Contributions: R.F.J. was responsible for study conception, analysis, write up, revision and final editing; Z.I. carried out the literature search, data analysis and write up; K.A., J.H. and N.A. were responsible for data collection; W.F. was responsible for final manuscript revision and approval. All authors have read and agreed to the published version of the manuscript.

Funding: This research received no funding.

Institutional Review Board Statement: No IRB approval was needed.

Informed Consent Statement: Not applicable.

Data Availability Statement: The data that support the findings of this study are available from the corresponding author upon reasonable request.

Acknowledgments: We would like to thank the Clinical Research Institute at AUBMC that supported us in data analysis.

Conflicts of Interest: The authors declare that they have no conflict of interest.

References

1. Karin, M. Nuclear factor-kappaB in cancer development and progression. *Nature* **2006**, *441*, 431–436. [CrossRef]
2. Singh, N.; Baby, D.; Rajguru, J.P.; Patil, P.B.; Thakkannavar, S.S.; Pujari, V.B. Inflammation and cancer. *Ann. Afr. Med.* **2019**, *18*, 121–126. [CrossRef]
3. Mantovani, A. Cancer: Inflammation by remote control. *Nature* **2005**, *435*, 752–753. [CrossRef]
4. Cordon-Cardo, C.; Prives, C. At the crossroads of inflammation and tumorigenesis. *J. Exp. Med.* **1999**, *190*, 1367–1370. [CrossRef] [PubMed]

5. Itzkowitz, S.H.; Yio, X. Inflammation and cancer IV. Colorectal cancer in inflammatory bowel disease: The role of inflammation. *Am. J. Physiol. Gastrointest. Liver Physiol.* **2004**, *287*, G7–G17. [CrossRef]
6. Peter, S.; Beglinger, C. Helicobacter pylori and gastric cancer: The causal relationship. *Digestion* **2007**, *75*, 25–35. [CrossRef]
7. Chin, A.I.; Dempsey, P.W.; Bruhn, K.; Miller, J.F.; Xu, Y.; Cheng, G. Involvement of receptor-interacting protein 2 in innate and adaptive immune responses. *Nature* **2002**, *416*, 190–194. [CrossRef]
8. Kobayashi, K.; Inohara, N.; Hernandez, L.D.; Galán, J.E.; Núñez, G.; Janeway, C.A.; Medzhitov, R.; Flavell, R.A. RICK/Rip2/CARDIAK mediates signalling for receptors of the innate and adaptive immune systems. *Nat. Cell Biol.* **2002**, *416*, 194–199. [CrossRef]
9. Cai, X.; Yang, Y.; Xia, W.; Kong, H.; Wang, M.; Fu, W.; Long, M.; Hu, Y.; Xu, D. RIP2 promotes glioma cell growth by regulating TRAF3 and activating the NF-κB and p38 signaling pathways. *Oncol. Rep.* **2018**, *39*, 2915–2923. [CrossRef]
10. Chan, L.-P.; Wang, L.-F.; Chiang, F.-Y.; Lee, K.-W.; Kuo, P.-L.; Liang, C.-H. IL-8 promotes HNSCC progression on CXCR1/2-meidated NOD1/RIP2 signaling pathway. *Oncotarget* **2016**, *7*, 61820–61831. [CrossRef]
11. Xu, Y.; Zhang, Z.; Zhang, L.; Zhang, C. Novel module and hub genes of distinctive breast cancer associated fibroblasts identified by weighted gene co-expression network analysis. *Breast Cancer* **2020**, *27*, 1017–1028. [CrossRef] [PubMed]
12. Singel, S.M.; Batten, K.; Cornelius, C.; Jia, G.; Fasciani, G.; Barron, S.L.; Wright, W.E.; Shay, J.W. Receptor-interacting protein kinase 2 promotes triple-negative breast cancer cell migration and invasion via activation of nuclear factor-kappaB and c-Jun N-terminal kinase pathways. *Breast Cancer Res.* **2014**, *16*, R28. [CrossRef] [PubMed]
13. Jaafar, R.; Mnich, K.; Dolan, S.; Hillis, J.; Almanza, A.; Logue, S.E.; Samali, A.; Gorman, A.M. RIP2 enhances cell survival by activation of NF-kB in triple negative breast cancer cells. *Biochem. Biophys. Res. Commun.* **2018**, *497*, 115–121. [CrossRef] [PubMed]
14. Zare, A.; Petrova, A.; Agoumi, M.; Armstrong, H.; Bigras, G.; Tonkin, K.; Wine, E.; Baksh, S. RIPK2: New Elements in Modulating Inflammatory Breast Cancer Pathogenesis. *Cancers* **2018**, *10*, 184. [CrossRef]
15. Guirado, M.; Gil, H.; Saenz-Lopez, P.; Reinboth, J.; Garrido, F.; Cozar, J.M.; Ruiz-Cabello, F.; Carretero, R. Association between C13ORF31, NOD2, RIPK2 and TLR10 polymorphisms and urothelial bladder cancer. *Hum. Immunol.* **2012**, *73*, 668–672. [CrossRef]
16. Yang, Q.; Tian, S.; Liu, Z.; Dong, W. Knockdown of RIPK2 Inhibits Proliferation and Migration, and Induces Apoptosis via the NF-κB Signaling Pathway. *Front. Genet.* **2021**. [CrossRef]
17. Yan, Y.; Zhou, B.; Qian, C.; Vasquez, A.; Chatterjee, A.; Yuan, X.; Posadas, E.; Kyprianou, N.; Knudsen, B.S.; Murali, R.; et al. RIPK2 Stabilizes c-Myc and is an Actionable Target for Inhibiting Prostate Cancer Metastasis. *bioRxiv* **2020**. [CrossRef]
18. Long, A.G.; Lundsmith, E.T.; Hamilton, K.E. Inflammation and Colorectal Cancer. *Curr. Color. Cancer Rep.* **2017**, *13*, 341–351. [CrossRef]
19. Slattery, M.L.; Mullany, L.E.; Sakoda, L.; Samowitz, W.S.; Wolff, R.K.; Stevens, J.R.; Herrick, J.S. The NF-κB signalling pathway in colorectal cancer: Associations between dysregulated gene and miRNA expression. *J. Cancer Res. Clin. Oncol.* **2018**, *144*, 269–283. [CrossRef]
20. Stronati, L.; Negroni, A.; Pierdomenico, M.; D'Ottavio, C.; Tirindelli, D.; Di Nardo, G.; Oliva, S.; Viola, F.; Cucchiara, S. Altered expression of innate immunity genes in different intestinal sites of children with ulcerative colitis. *Dig. Liver Dis.* **2010**, *42*, 848–853. [CrossRef]
21. Couturier-Maillard, A.; Secher, T.; Rehman, A.; Normand, S.; De Arcangelis, A.; Häsler, R.; Huot, L.; Grandjean, T.; Bressenot, A.; Delanoye-Crespin, A.; et al. NOD2-mediated dysbiosis predisposes mice to transmissible colitis and colorectal cancer. *J. Clin. Investig.* **2013**, *123*, 700–711. [CrossRef] [PubMed]
22. Flebbe, H.; Hamdan, F.H.; Kari, V.; Kitz, J.; Gaedcke, J.; Ghadimi, B.M.; Johnsen, S.A.; Grade, M. Epigenome Mapping Identifies Tumor-Specific Gene Expression in Primary Rectal Cancer. *Cancers* **2019**, *11*, 1142. [CrossRef] [PubMed]
23. Luo, Z.-W.; Zhu, M.-G.; Zhang, Z.-Q.; Ye, F.-J.; Huang, W.-H.; Luo, X.-Z. Increased expression of Ki-67 is a poor prognostic marker for colorectal cancer patients: A meta analysis. *BMC Cancer* **2019**, *19*, 123. [CrossRef] [PubMed]
24. Morikawa, T.; Kuchiba, A.; Liao, X.; Imamura, Y.; Yamauchi, M.; Qian, Z.R.; Nishihara, R.; Sato, K.; Meyerhardt, J.A.; Fuchs, C.S.; et al. Tumor TP53 expression status, body mass index and prognosis in colorectal cancer. *Int. J. Cancer* **2011**, *131*, 1169–1178. [CrossRef]
25. Wang, P.; Liang, J.; Wang, Z.; Hou, H.; Shi, L.; Zhou, Z. The prognostic value of p53 positive in colorectal cancer: A retrospective cohort study. *Tumor Biol.* **2017**, *39*, 1010428317703651. [CrossRef] [PubMed]
26. Araújo, R.F.; Lira, G.A.; Vilaça, J.A.; Guedes, H.G.; Leitão, M.C.A.; Lucena, H.F.; Ramos, C. Prognostic and diagnostic implications of MMP-2, MMP-9, and VEGF-α expressions in colorectal cancer. *Pathol. Res. Pract.* **2015**, *211*, 71–77. [CrossRef]
27. Bendardaf, R.; Buhmeida, A.; Hilska, M.; Laato, M.; Syrjänen, S.; Syrjänen, K.; Collan, Y.; Pyrhönen, S. VEGF-1 expression in colorectal cancer is associated with disease localization, stage, and long-term disease-specific survival. *Anticancer Res.* **2009**, *28*, 3865–3870.
28. Mohamed, A.H.; Said, N.M. Immunohistochemical Expression of Fatty Acid Synthase and Vascular Endothelial Growth Factor in Primary Colorectal Cancer: A Clinicopathological Study. *J. Gastrointest. Cancer* **2018**, *50*, 485–492. [CrossRef]
29. Mizuno, H.; Kitada, K.; Nakai, K.; Sarai, A. PrognoScan: A new database for meta-analysis of the prognostic value of genes. *BMC Med. Genom.* **2009**, *2*, 18. [CrossRef]
30. Lea, D.; Håland, S.; Hagland, H.R.; Søreide, K. Accuracy of TNM staging in colorectal cancer: A review of current culprits, the modern role of morphology and stepping-stones for improvements in the molecular era. *Scand. J. Gastroenterol.* **2014**, *49*, 1153–1163. [CrossRef]

31. Bray, F.; Ferlay, J.; Soerjomataram, I.; Siegel, R.L.; Torre, L.A.; Jemal, A. Global cancer statistics 2018: GLOBOCAN estimates of incidence and mortality worldwide for 36 cancers in 185 countries. *CA Cancer J. Clin.* **2018**, *68*, 394–424. [CrossRef] [PubMed]
32. Weitz, J.; Koch, M.; Debus, J.; Höhler, T.; Galle, P.R.; Büchler, M.W. Colorectal cancer. *Lancet* **2005**, *365*, 153–165. [CrossRef]
33. Piawah, S.; Venook, A.P. Targeted therapy for colorectal cancer metastases: A review of current methods of molecularly targeted therapy and the use of tumor biomarkers in the treatment of metastatic colorectal cancer. *Cancer* **2019**, *125*, 4139–4147. [CrossRef]
34. Van Cutsem, E.; Köhne, C.H.; Hitre, E.; Zaluski, J.; Chang Chien, C.R.; Makhson, A.; D'Haens, G.; Pintér, T.; Lim, R.; Bodoky, G.; et al. Cetuximab and chemotherapy as initial treatment for metastatic colorectal cancer. *N. Engl. J. Med.* **2009**, *360*, 1408–1417. [CrossRef] [PubMed]
35. Cao, D.-D.; Xu, H.-L.; Xu, X.-M.; Ge, W. The impact of primary tumor location on efficacy of cetuximab in metastatic colorectal cancer patients with different Kras status: A systematic review and meta-analysis. *Oncotarget* **2017**, *8*, 53631–53641. [CrossRef]
36. Wu, S.; Kanda, T.; Nakamoto, S.; Imazeki, F.; Yokosuka, O. Knockdown of receptor-interacting serine/threonine protein kinase-2 (RIPK2) affects EMT-associated gene expression in human hepatoma cells. *Anticancer Res.* **2012**, *32*, 3775–3783.
37. Jensen, S.; Seidelin, J.B.; Lacasse, E.C.; Nielsen, O.H. SMAC mimetics and RIPK inhibitors as therapeutics for chronic inflammatory diseases. *Sci. Signal.* **2020**, *13*, eaax8295. [CrossRef]

MDPI
St. Alban-Anlage 66
4052 Basel
Switzerland
www.mdpi.com

Medicina Editorial Office
E-mail: medicina@mdpi.com
www.mdpi.com/journal/medicina

www.ingramcontent.com/pod-product-compliance
Lightning Source LLC
Chambersburg PA
CBHW040717270926
51850CB00023B/2774
* 9 7 8 3 0 3 6 5 9 4 4 1 5 *